MONOGRAPHS ON STATISTICS AND APPLIED PROBABILITY

General Editors

D.R. Cox, D.V. Hinkley, N. Reid, D.B. Rubin and B.W. Silverman

1 Stochastic Population Models in Ecology and Epidemology *M.S. Bartlett* (1960)
2 Queues *D.R. Cox and W.L. Smith* (1961)
3 Monte Carlo Methods *J.M. Hammersley and D.C. Handscomb* (1964)
4 The Statistical Analysis of Series of Events *D.R. Cox and P.A.W. Lewis* (1966)
5 Population Genetics *W.J. Ewens* (1969)
6 Probability, Statistics and Time *M.S. Barlett* (1975)
7 Statistical Inference *S.D. Slivey* (1975)
8 The Analysis of Contingency Tables *B.S. Everitt* (1977)
9 Multivariate Analysis in Behavioural Research *A.E. Maxwell* (1977)
10 Stochastic Abundance Models *S. Engen* (1978)
11 Some Basic Theory for Statistical Inference *E.J.G. Pitman* (1978)
12 Point Processes *D.R. Cox and V. Isham* (1980)
13 Identification of Outliers *D.M. Hawkins* (1980)
14 Optimal Design *S.D. Silvey* (1980)
15 Finite Mixture Distributions *B.S. Everitt and D.J. Hand* (1981)
16 Classification *A.D. Gordon* (1981)
17 Distribution-free Statistical Methods *J.S. Maritz* (1981)
18 Residuals and Influence in Regression *R.D. Cook and S. Weisberg* (1982)
19 Applications of Queueing Theory *G.F. Newell* (1982)
20 Risk Theory, 3rd edition *R.E. Beard, T. Pentikainen and E. Pesonen* (1984)
21 Analysis of Survival Data *D.R. Cox and D. Oakes* (1984)
22 An Introduction to Latent Variable Models *B.S. Everitt* (1984)
23 Bandit Problems *D.A. Berry and B. Fristedt* (1985)
24 Stochastic Modelling and Control *M.H.A. Davis and R. Vinter* (1985)
25 The Statistical Analysis of Compositional Data *J. Aitchison* (1986)
26 Density Estimation for Statistical and Data Analysis *B.W. Silverman*
27 Regression Analysis with Application *G.B. Wetherill* (1986)
28 Sequential Methods in Statistics, 3rd edition *G.B. Wetherill* (1986)
29 Tensor Methods in Statistics *P. McCullagh* (1987)
30 Transformation and Weighting in Regression *R.J. Carroll and D. Ruppert* (1988)

31 Asymptotic Techniques for Use in Statistics *O.E. Barndorff-Nielson and D.R. Cox* (1989)

32 Analysis of Binary Data, 2nd edition *D.R. Cox and E.J. Snell* (1989)

33 Analysis of Infectious Disease Data *N.G. Becker* (1989)

34 Design and Analysis of Cross-Over Trials *B. Jones and M.G. Kenward* (1989)

35 Empirical Bayes Method, 2nd edition *J.S. Maritz and T. Lwin* (1989)

36 Symmetric Multivariate and Related Distributions *K.-T. Fang S. Kotz and K. Ng* (1989)

37 Generalized Linear Models, 2nd edition *P. McCullagh and J.A. Nelder* (1989)

38 Cyclic Designs *J.A. John* (1987)

39 Analog Estimation Mehtods in Econometrics *C.F. Manski* (1988)

40 Subset Selection in Regression *A.J. Miller* (1990)

41 Analysis of Repeated Measures *M. Crowder and D.J. Hand* (1990)

42 Statistical Reasoning with Imprecise Probabilities *P. Walley* (1990)

43 Generalized Additive Models *T.J. Hastie and R.J. Tibshirani* (1990)

44 Inspection Errors for Attributes in Quality Control *N.L. Johnson, S. Kotz and X. Wu* (1991)

45 The Analysis of Contingency Tables, 2nd edition *B.S. Everitt* (1992)

47 Longitudinal Data with Serial Correlation: A State-space Approach *R.H. Jones* (1993)

48 Differential Geometry and Statistics *M.K. Murray and J.W. Rice* (1993)

49 Markov Models and Optimization *M.H.A. Davis* (1993)

50 Networks and Chaos – Statistical and Probalistic Aspects *O.E. Barndorff-Nielson, J.H. Jensen and W.S. Kendall* (1993)

51 Number Theoretic Methods in Statistics *K.-T. Fang and Y. Wang* (1993)

52 Inference and Asymptotics *D.R. Cox and O.E. Barndorff-Nielson* (1994)

53 Practical Risk Theory for Actuaries *C.D. Daykin, T. Pentikäinen and M. Pesonen* (1993)

54 Statistical Concepts and Applications in Medicine *J. Aitchison and I.J. Lauder* (1994)

55 Predictive Inference *S. Geisser* (1993)

56 Model Free Curve Estimation *M. Tartar and M. Lock* (1994)

57 An Introduction to the Bootstrap *B. Efron and R. Tibshirani* (1993)

58 Nonparametric Regression and Generalized Linear Models *P.J. Green and B.W. Silverman* (1994)

(Full details concerning this series are available from the publisher)

Nonparametric Regression and Generalized Linear Models

A ROUGHNESS PENALTY APPROACH

P.J. GREEN

and

B.W. SILVERMAN

School of Mathematics
University of Bristol
UK

CHAPMAN & HALL

London · Glasgow · New York · Tokyo · Melbourne · Madras

Published by Chapman & Hall, 2-6 Boundary Row, London SE1 8HN, UK

Chapman & Hall, 2-6 Boundary Row, London SE1 8HN, UK

Blackie Academic & Professional, Wester Cleddens Road, Bishopbriggs, Glasgow G64 2NZ, UK

Chapman & Hall Inc., One Penn Plaza, 41st Floor, New York, NY10119, USA

Chapman & Hall Japan, Thomson Publishing Japan, Hirakawacho Nemoto Building, 6F, 1-7-11 Hirakawa-cho, Chiyoda-ku, Tokyo 102, Japan

Chapman & Hall Australia, Thomas Nelson Australia, 102 Dodds Street, South Melbourne, Victoria 3205, Australia

Chapman & Hall India, R. Seshadri, 32 Second Main Road, CIT East, Madras 600 035, India

First edition 1994

Printed in England by Clays Ltd, St Ives Plc, Bungay, Suffolk

ISBN 0 412 30040 0

A catalogue record for this book is available from the British Library

Library of Congress Cataloging-in-Publication data available

Contents

Preface

In the last fifteen years there has been an upsurge of interest and activity in the general area of nonparametric smoothing in statistics. Many methods have been proposed and studied. Some of the most popular of these are primarily 'data-analytic' in flavour and do not make particular use of statistical models. The roughness penalty approach, on the other hand, provides a 'bridge' towards classical and parametric statistics. It allows smoothing to be incorporated in a natural way not only into regression but also much more generally, for example into problems approached by generalized linear modelling. In this monograph we have tried to convey our personal view of the roughness penalty method, and to show how it provides a unifying approach to a wide range of smoothing problems.

We hope that the book will be of interest both to those coming to the area for the first time and to readers more familiar with the field. While advanced mathematical ideas have been valuable in some of the theoretical development, the methodological power of roughness penalty methods can be demonstrated and discussed without the need for highly technical mathematical concepts, and as far as possible we have aimed to provide a self-contained treatment that depends only on simple linear algebra and calculus.

For various ways in which they have helped us, we would like to thank Tim Cole, John Gavin, Trevor Hastie, Guy Nason, Doug Nychka, Christine Osborne, Glenn Stone, Rob Tibshirani and Brian Yandell. Preliminary versions of parts of the book have been used as the basis of courses given to successive classes of MSc students at the University of Bath, and to graduate students at Stanford University and at the University of São Paulo; in all cases the feedback and reactions have been most helpful. The book was mainly written while one of us (BWS) held a post at the University of Bath, and he would like to pay grateful tribute to the intellectual and material environment provided there.

Peter Green and Bernard Silverman
Bristol, September 1993

CHAPTER 1

Introduction

1.1 Approaches to regression

The main theme of this book is the application of the *roughness penalty* approach to problems in regression and related fields. Before going on to introduce roughness penalties in Section 1.2, it is helpful to set the scene by briefly discussing linear regression first of all.

1.1.1 Linear regression

Linear regression is one of the oldest and most widely used statistical techniques. Given data pairs $(t_i, Y_i), i = 1, \ldots, n$, the natural way to view linear regression is as a method fitting a *model* of the form

$$Y = a + bt + \text{ error} \tag{1.1}$$

to the observed data. Of course, linear regression is often applied rather blindly to data without any particular aim in mind. It is helpful, however, to identify two of the main purposes for which linear regression is useful. The distinction between these is not at all rigid and often both of them will apply.

The first main purpose of regression is to provide a summary or reduction of the observed data in order to *explore* and *present* the relationship between the design variable t and the response variable Y. It is obvious and natural, when given a plot of data displaying an approximate linear trend, to draw a straight line to emphasize this trend. Linear regression automates this procedure and ensures comparability and consistency of results.

The other main purpose of regression is to use the model (1.1) for *prediction*; given any point t, an estimate of the expected value of a new observation Y at the point t is given by $\hat{a} + \hat{b}t$, where $\hat{a}$ and $\hat{b}$ are estimates of a and b. Prediction is the usual context in which elementary textbooks introduce the idea of linear regression. While prediction is undoubtedly an important aspect of regression, it is probably a much more accurate reflection of statistical practice to consider regression primarily as a *model-based* method for data summary.

The model-based foundation of regression distinguishes it somewhat from the more exclusively data-based ideas underlying more purely exploratory techniques such as those presented by Tukey (1977). Adopting an approach based on a model has both advantages and disadvantages. One advantage is that the methodology can be extended to a wide range of alternative data structures, for example along the lines of generalized linear models as discussed by McCullagh and Nelder (1989).

1.1.2 Polynomial regression

There are very many data sets where it is clearly inappropriate to fit a straight line model of the form (1.1) and where a model of the form

$$y = g(t) + \text{ error} \tag{1.2}$$

is called for, where g is a curve of some sort. The classical approach is to use for g a low order polynomial, the coefficients of which are estimated by least squares. This approach is widely used in practice and is easily implemented using a multiple regression approach.

Polynomial regression is a popular technique but it does suffer from various drawbacks. One of these is that individual observations can exert an influence, in unexpected ways, on remote parts of the curve. Another difficulty is that the model elaboration implicit in increasing the polynomial degree happens in discrete steps and cannot be controlled continuously. A third point, not necessarily a drawback, is that polynomials of a given degree form a 'hard-edged' finite-dimensional class of models and that there may be some advantage in allowing the data to determine the fitted model in a somewhat more flexible way.

1.2 Roughness penalties

1.2.1 The aims of curve fitting

In its simplest form the roughness penalty approach is a method for relaxing the model assumptions in classical linear regression along lines a little different from polynomial regression.

Consider, first, what would happen if we were to attempt to fit a model of the form (1.2) by least squares, without placing any restrictions on the curve g. It is then, of course, the case that the residual sum of squares can be reduced to zero by choosing g to interpolate the given data; for example one could join the given points to obtain the function g shown

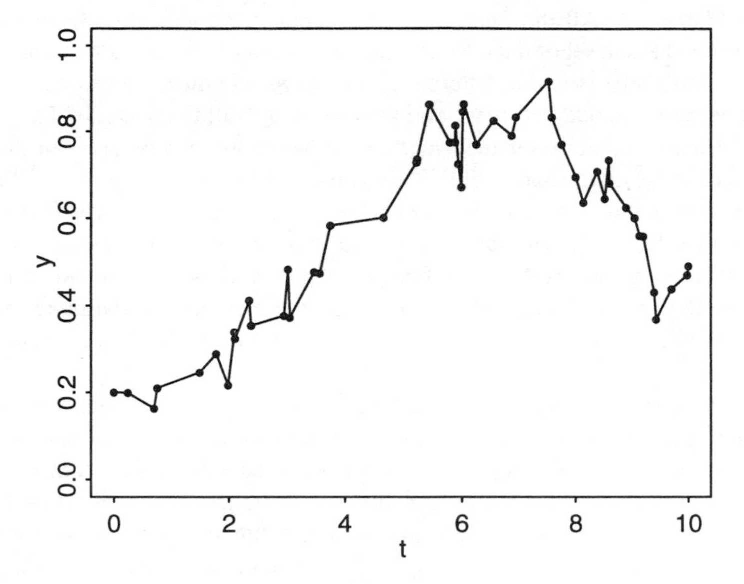

Figure 1.1. *Synthetic data joined by straight lines.*

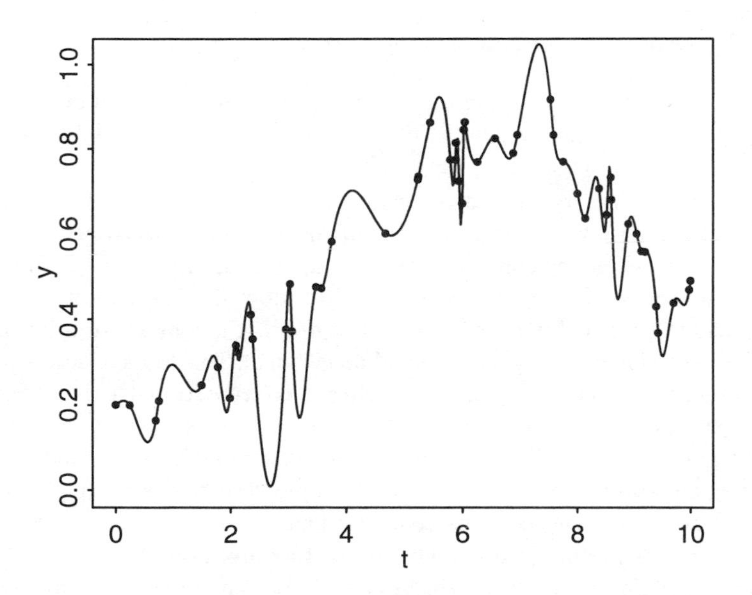

Figure 1.2. *Synthetic data interpolated by a curve with continuous second derivative.*

in Figure 1.1. All the figures in this section are constructed from the same synthetic set of data. Placing smoothness conditions on the g does not make any essential difference; the curve shown in Figure 1.2 has continuous second derivative and passes through all the points (t_i, Y_i).

What is it that makes the functions shown in these two curves unsatisfactory as explanations of the given data? At the outset, it should be stressed that in some situations they are not at all unsatisfactory. It may well be that the phenomenon under study is known to vary rapidly and that the given observations are known to be extremely accurate; however even in this case it is of interest to regard the very local variation in the curve as random 'noise' in order to study the more slowly varying 'trend' in the data.

The curves in Figures 1.1 and 1.2 underline the point that a good fit to the data is not the one and only aim in curve fitting; another, often conflicting, aim is to obtain a curve estimate that does not display too much rapid fluctuation. The basic idea of the roughness penalty approach is to quantify the notion of a rapidly fluctuating curve and then to pose the estimation problem in a way that makes explicit the necessary compromise between the two rather different aims in curve estimation.

1.2.2 Quantifying the roughness of a curve

Given a curve g defined on an interval $[a, b]$, there are many different ways of measuring how 'rough' or 'wiggly' the curve g is. An intuitively appealing way of measuring the roughness of a twice-differentiable curve g is to calculate its integrated squared second derivative $\int_a^b \{g''(t)\}^2 dt$. There is a variety of ways of motivating this measure of roughness.

Particularly in the context of regression, it is natural for any measure of roughness not to be affected by the addition of a constant or linear function, so that if two functions differ only by a constant or a linear function then their roughness should be identical. This leads naturally to the idea of a roughness functional that depends on the second derivative of the curve under consideration. Of course, one could conceivably consider the maximum of $|g''|$ or the number of inflection points in g, but the integrated squared second derivative is a global measure of roughness that has, as we shall see, considerable computational advantages.

One attractive motivation arises from a formalization of a mechanical device that was often used (in the age before computer graphics) for drawing smooth curves. If a thin piece of flexible wood, called a *spline*, is bent to the shape of the graph of g then the leading term in the strain energy is proportional to $\int g''^2$.

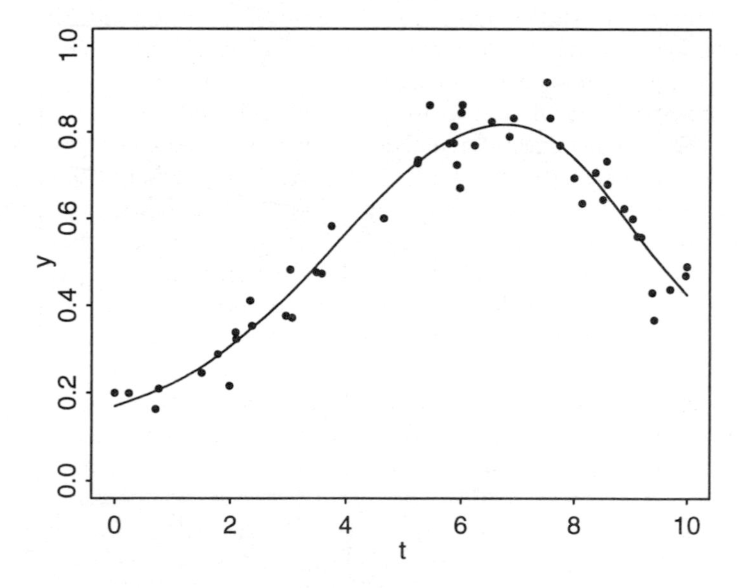

Figure 1.3. *Synthetic data with the curve that minimizes $S(g)$ with $\alpha = 1$.*

1.2.3 Penalized least squares regression

The roughness penalty approach to curve estimation is now easily stated. Given any twice-differentiable function g defined on $[a, b]$, and a smoothing parameter $\alpha > 0$, define the penalized sum of squares

$$S(g) = \sum_{i=1}^{n} \{Y_i - g(t_i)\}^2 + \alpha \int_a^b \{g''(x)\}^2 dx. \tag{1.3}$$

The penalized least squares estimator $\hat{g}$ is defined to be the minimizer of the functional $S(g)$ over the class of all twice-differentiable functions g. In Chapter 2 we shall explain how $\hat{g}$ can be characterized and computed. Some remarks about particular software implementations are made in Chapter 8.

The addition of the roughness penalty term $\alpha \int g''^2$ in (1.3) ensures that the cost $S(g)$ of a particular curve is determined not only by its goodness-of-fit to the data as quantified by the residual sum of squares $\sum\{Y_i - g(t_i)\}^2$ but also by its roughness $\int g''^2$. The smoothing parameter α represents the 'rate of exchange' between residual error and local variation and gives the amount in terms of summed square residual error that corresponds to one unit of integrated squared second derivative. For the given value of α, minimizing $S(g)$ will give the best compromise

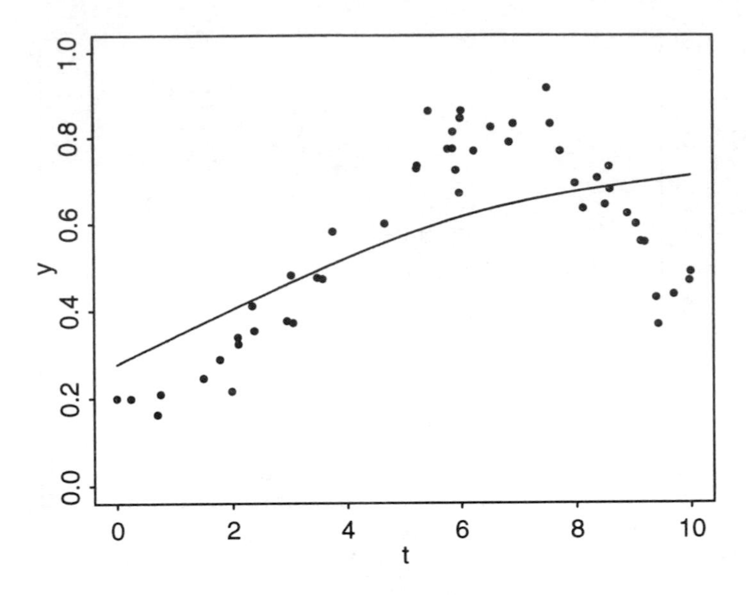

Figure 1.4. *Synthetic data with the curve that minimizes $S(g)$ for a large value of α.*

between smoothness and goodness-of-fit. An example of this approach applied to the synthetic data of Figure 1.1 is given in Figure 1.3.

If α is large then the main component in $S(g)$ will be the roughness penalty term and hence the minimizer $\hat{g}$ will display very little curvature. An example is given in Figure 1.4; in the limiting case as α tends to infinity the term $\int \hat{g}''^2$ will be forced to zero and the curve $\hat{g}$ will approach the linear regression fit.

On the other hand if α is relatively small then the main contribution to $S(g)$ will be the residual sum of squares, and the curve estimate $\hat{g}$ will track the data closely even if it is at the expense of being rather variable. For an illustration see Figure 1.5. In the limit as α tends to zero, $\hat{g}$ will approach the interpolating curve shown in Figure 1.2. The whole question of how to choose the value of α most appropriate to a given data set is of course an important one which will be addressed in Chapter 3. Many other aspects of the roughness penalty approach to one-dimensional regression smoothing are discussed in Chapters 2 and 3.

One of the reasons that roughness penalty methods are not as widely taught or used as they might be is the mistaken impression that they require deep technical mathematical knowledge for their development and

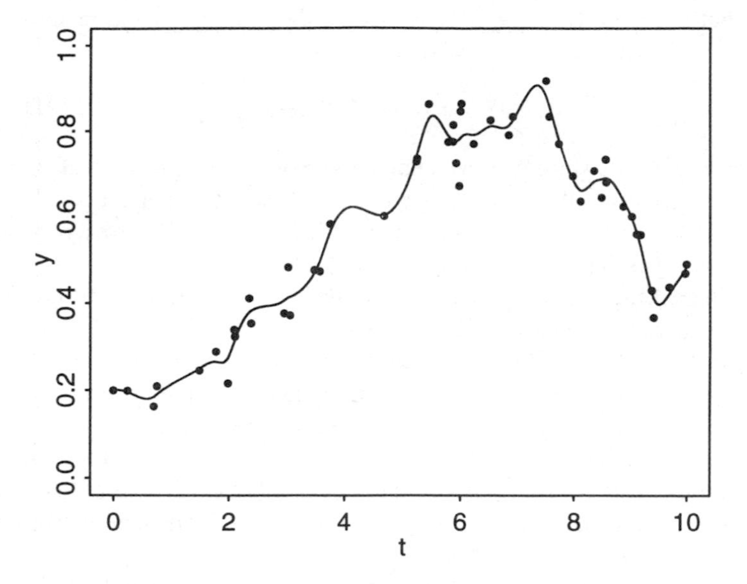

Figure 1.5. *Synthetic data with the curve that minimizes $S(g)$ for a small value of α.*

understanding. In fact all that is required to deal with the mathematical background to the practical aspects of the methods discussed in this book is some elementary calculus and numerical linear algebra. In some of the more advanced technical literature there is much mention of concepts such as Sobolev spaces, reproducing kernels, Hilbert spaces and the calculus of variations, but we shall not be mentioning any of these explicitly, nor shall we be discussing any detail of the asymptotic theory of the estimators.

1.3 Extensions of the roughness penalty approach

Our main theme is the applicability of the roughness penalty approach in a wide variety of contexts. For example, in Chapter 4 we discuss *semiparametric* modelling, a simple application of roughness penalties to multiple regression. Suppose we observe a variable Y that depends on a multivariate explanatory variable. In multiple linear regression it would be assumed that this dependence was linear, and the familiar theory of the general linear model would be used to estimate it. The idea of semiparametric models, in their simplest form, is to relax the assumption of linearity on just one of the explanatory variables, which we shall call

t, but to retain the linear dependence on the vector $\mathbf{x}$ of the remaining variables. This yields a model of the form

$$Y = g(t) + \mathbf{x}^T\beta + \text{ error,} \tag{1.4}$$

where $g(t)$ is a smooth curve and β a vector of parameters, both to be estimated. As well as describing the way in which the roughness penalty approach can be applied to semiparametric modelling, we give a number of examples that illustrate the method.

One of the most important developments in statistics in the last 25 years has been the introduction and widespread use of *generalized linear models* (GLMs), as formulated by Nelder and Wedderburn (1972). In Chapter 5 we review the GLM structure, and explain how a roughness penalty approach allows some of the linear dependences in Nelder and Wedderburn's structure to be relaxed to dependence on variables through smooth curves g. In Chapter 6, we discuss further extensions both to the basic one-dimensional smoothing method, and to the methodology based on generalized linear models.

The natural extension of the model (1.4) is to allow the dependence on *all* the explanatory variables to be nonlinear. Suppose that $\mathbf{t}$ is a d-dimensional vector of explanatory variables $t_1, \ldots, t_d$. The full extension would be to a model of the form

$$Y = g(\mathbf{t}) + \text{ error,} \tag{1.5}$$

where g is a d-dimensional surface. The estimation of g by a roughness penalty method is discussed in Chapter 7, with particular emphasis on the two-dimensional case.

Mention should also be made here of the *additive model* approach set out in the monograph by Hastie and Tibshirani (1990). This can be viewed as a 'half-way house' between multiple linear regression and the full surface-fitting approach implied by (1.5). It is assumed that the mean response is a sum of functions of the individual explanatory variables,

$$Y = \sum_{j=1}^{d} g_j(t_j) + \text{ error,}$$

where all or some of the d functions g_j are smooth curves, and the remainder, if any, are assumed to be linear. We shall discuss the additive model approach and its ramifications for GLMs briefly in Sections 4.8 and 5.6, but for full details we refer the reader to Hastie and Tibshirani (1990).

1.4 Computing the estimates

It is our hope that readers will be stimulated to apply the methods described in this book to their own data. In principle, we have given enough information for readers to write their own programs, and would recommend students (and others) to produce their own implementations as a valuable exercise. However, the existence of publicly-available software is of course essential for the widespread application of statistical techniques. Because of the variety and rapid development of computing facilities, we have discussed specific software implementations separately, in Chapter 8. We describe routines available within the statistical language S (Becker, Chambers and Wilks, 1988) and the FORTRAN package GCVPACK of Bates, Lindstrom, Wahba and Yandell (1987).

1.5 Further reading

In this book we have not attempted to survey the subject of roughness penalty methods exhaustively, because we believe that in many ways a 'personal view' is more useful. Other recent books and monographs which provide a variety of treatments of roughness penalty methods in particular, and of statistical smoothing methods more generally, include Eubank (1988), Wahba (1990), Härdle (1990), Hastie and Tibshirani (1990) and Rosenblatt (1991).

CHAPTER 2

Interpolating and smoothing splines

2.1 Cubic splines

In this chapter we explain how the curve $\hat{g}$ that minimizes the penalized sum of squares (1.3) can be found. A pivotal rôle in our discussion is played by *cubic splines*. We shall first describe what a cubic spline is, and then explain how cubic splines arise in interpolation and nonparametric regression. Some algorithmic material will also be presented.

There is an enormous literature on splines, most of it concerning their numerical-analytic rather than statistical properties. Books on splines include, for example, Ahlberg, Nilson and Walsh (1967), Prenter (1975), De Boor (1978) and Schumaker (1993). Although the notation is at times somewhat cumbersome, we reiterate that our treatment requires no mathematics beyond simple linear algebra and calculus.

2.1.1 What is a cubic spline?

Suppose we are given real numbers $t_1, \ldots, t_n$ on some interval $[a, b]$, satisfying $a < t_1 < t_2 < \ldots < t_n < b$. A function g defined on $[a, b]$ is a *cubic spline* if two conditions are satisfied. Firstly, on each of the intervals $(a, t_1), (t_1, t_2), (t_2, t_3), \ldots, (t_n, b)$, g is a cubic polynomial; secondly the polynomial pieces fit together at the points t_i in such a way that g itself and its first and second derivatives are continuous at each t_i, and hence on the whole of $[a, b]$.

The points t_i are called *knots*. There are many essentially equivalent ways of specifying a cubic spline. One obvious way is to give the four polynomial coefficients of each cubic piece, for example in the form

$$g(t) = d_i(t - t_i)^3 + c_i(t - t_i)^2 + b_i(t - t_i) + a_i \text{ for } t_i \leq t \leq t_{i+1} \qquad (2.1)$$

for given constants $a_i, b_i, c_i, d_i, i = 0, \ldots, n$; we define $t_0 = a$ and $t_{n+1} = b$.

The continuity conditions on g and on its first two derivatives imply various relations between the coefficients. For example, the continuity of

g at t_{i+1} yields, for $i = 0, \ldots, n-1$,

$$d_i(t_{i+1} - t_i)^3 + c_i(t_{i+1} - t_i)^2 + b_i(t_{i+1} - t_i) + a_i = a_{i+1} \tag{2.2}$$

since both expressions are equal to $g(t_{i+1})$.

A cubic spline on an interval $[a, b]$ will be said to be a *natural cubic spline* (NCS) if its second and third derivatives are zero at a and b. These conditions are called the *natural boundary conditions*. They imply that $d_0 = c_0 = d_n = c_n = 0$, so that g is linear on the two extreme intervals $[a, t_1]$ and $[t_n, b]$.

2.1.2 The value-second derivative representation

In fact it turns out that (2.1) is not the most convenient representation of a natural cubic spline either for computation or for mathematical discussion. We shall instead specify a NCS by giving its value and second derivative at each of the knots t_i. This representation will be called the *value-second derivative representation*. Suppose that g is a NCS with knots $t_1 < \ldots < t_n$. Define

$$g_i = g(t_i) \text{ and } \gamma_i = g''(t_i) \text{ for } i = 1, \ldots, n.$$

By the definition of a NCS the second derivative of g at t_1 and at t_n is zero, so that $\gamma_1 = \gamma_n = 0$. Let $\mathbf{g}$ be the vector $(g_1, \ldots, g_n)^T$ and let $\boldsymbol{\gamma}$ be the vector $(\gamma_2, \ldots, \gamma_{n-1})^T$. Note that the entries γ_i of the $(n-2)$-vector $\boldsymbol{\gamma}$ are numbered in a non-standard way, starting at $i = 2$; this will make for considerable simplification later on.

The vectors $\mathbf{g}$ and $\boldsymbol{\gamma}$ specify the curve g completely, and it is possible to give explicit formulae in terms of $\mathbf{g}$ and $\boldsymbol{\gamma}$ for the value and derivatives of g at any point t. This enables g to be plotted to any desired degree of accuracy. Details of these formulae will be given in Section 2.4 below.

It turns out to be the case that not all possible vectors $\mathbf{g}$ and $\boldsymbol{\gamma}$ represent bona fide natural cubic splines. We shall now discuss a necessary and sufficient condition for the vectors genuinely to represent a natural cubic spline on the given knot sequence.

The condition depends on two band matrices Q and R which we now define. Let $h_i = t_{i+1} - t_i$ for $i = 1, \ldots, n-1$. Let Q be the $n \times (n-2)$ matrix with entries q_{ij}, for $i = 1, \ldots, n$ and $j = 2, \ldots, n-1$, given by

$$q_{j-1,j} = h_{j-1}^{-1}, \; q_{jj} = -h_{j-1}^{-1} - h_j^{-1}, \text{ and } q_{j+1,j} = h_j^{-1}$$

for $j = 2, \ldots, n-1$, and $q_{ij} = 0$ for $|i - j| \geq 2$. The columns of Q are numbered in the same non-standard way as the entries of $\boldsymbol{\gamma}$, starting at $j = 2$, so that the top left element of Q is q_{12}.

The symmetric matrix R is $(n-2) \times (n-2)$ with elements r_{ij}, for i and j running from 2 to $(n-1)$, given by

$$r_{ii} = \tfrac{1}{3}(h_{i-1} + h_i) \text{ for } i = 2, \dots, n-1,$$

$$r_{i,i+1} = r_{i+1,i} = \tfrac{1}{6}h_i \text{ for } i = 2, \dots, n-2,$$

and $r_{ij} = 0$ for $|i-j| \geq 2$.

The matrix R is strictly diagonal dominant, in the sense that $|r_{ii}| > \sum_{j \neq i} |r_{ij}|$ for each i. Standard arguments in numerical linear algebra (e.g. Todd (1962), Section 8.19) show that R is strictly positive-definite. We can therefore define a matrix K by

$$K = QR^{-1}Q^T. \tag{2.3}$$

The key property can now be stated, together with two important results that will be used later.

Theorem 2.1 *The vectors* $\mathbf{g}$ *and* γ *specify a natural cubic spline* g *if and only if the condition*

$$Q^T\mathbf{g} = R\gamma \tag{2.4}$$

is satisfied. If (2.4) is satisfied then the roughness penalty will satisfy

$$\int_a^b g''(t)^2 dt = \gamma^T R \gamma = \mathbf{g}^T K \mathbf{g}. \tag{2.5}$$

The proof of the theorem, for readers interested in the details, is given in Section 2.5 below.

2.2 Interpolating splines

Although our main emphasis in this book is on smoothing problems, it is helpful to spend a little time on the closely related problem of *interpolation*. The subject of interpolation is perhaps more familiar to numerical analysts than to statisticians, and our primary reason for introducing it here is to simplify and clarify the subsequent discussion of the smoothing problem. Nevertheless, the interpolation problem is of course of enormous importance in its own right.

Suppose we are given values $z_1, \dots, z_n$ at the points $t_1, \dots, t_n$. We wish to find a smooth curve g such that g *interpolates* the points (t_i, z_i), that is to say $g(t_i) = z_i$ for all $i = 1, \dots, n$. Obviously there are many ways of constructing a sensible interpolating function g. The simplest, and probably the most widely used, approach would be to join the given points (t_i, z_i) by straight lines. Whilst this undoubtedly suffices for many purposes, it does not yield a smooth curve, since the resulting function g has discontinuous derivative at each data point. The objections to this piecewise linear function are not just aesthetic; if the points are taken from a true underlying smooth curve, then it can be shown mathematically that a suitably chosen smooth interpolant will do a much better job of approximating the true underlying curve than will the piecewise linear interpolant.

Of course, just restricting attention to smooth interpolants does not give a unique answer. In a technical drawing office, smooth curves through given points are often drawn using *French curves*, which work by joining pieces of smooth curves together smoothly at points chosen subjectively in the light of the data. The success of this approach depends largely on the skill of the draftsman. Particularly because of the subjective nature of the choice both of the joins between the curves and of the curves themselves, it is a method that is not easy to formalize or automate.

An approach that is easier to define mathematically can be developed from our definition of roughness penalties in Section 1.2 above. Let $\mathcal{S}[a,b]$ be the space of all functions g on $[a,b]$ that have two continuous derivatives, and call a function *smooth* if it is in $\mathcal{S}[a,b]$. If we wanted the 'smoothest possible' curve that interpolated the given points, then a natural choice would be to use as our interpolant the curve that had the minimum value of $\int g''^2$ among all smooth curves that interpolate the data.

It turns out that among all curves g in $\mathcal{S}[a,b]$ interpolating the points (t_i, z_i), the one minimizing $\int g''^2$ is a natural cubic spline with knots t_i. Furthermore, provided $n \geq 2$, there is exactly one such natural cubic spline interpolating the data. Thus the problem of finding the interpolant with minimal $\int g''^2$ is precisely that of finding the unique natural cubic spline that has knots at the points t_i and values $g(t_i) = z_i$ for all i. We shall prove these assertions below and also demonstrate how the natural spline interpolant can be found by solving a system of linear equations.

The natural cubic spline interpolant has a mechanical motivation already alluded to in Section 1.2.2 above. Suppose a thin piece of flexible wood, a mechanical spline, is constrained to pass through the given points (t_i, z_i) but is otherwise free to fall into any shape. Mechanical splines are rather unusual nowadays, but were once in common use particularly for laying out the hulls of ships and for planning railway lines. In practice the mechanical spline is equipped with a number of pivoted sliding weights, called *ducks*. Placing the ducks at the points (t_i, z_i) on the drawing board will constrain the spline in the necessary way, and the spline will take up a position of minimum energy subject to the constraints. Provided the points lie reasonably close to a straight line, the spline's position will, to first order, describe a curve g minimizing $\int g''^2$ over curves interpolating the data.

2.2.1 Constructing the interpolating natural cubic spline

The result that it is always possible to interpolate a given set of values by a natural cubic spline, in a unique way, is so important that we shall give it as a theorem.

Theorem 2.2 *Suppose $n \geq 2$ and that $t_1 < \ldots < t_n$. Given any values $z_1, \ldots, z_n$, there is a unique natural cubic spline g with knots at the points t_i satisfying*

$$g(t_i) = z_i \text{ for } i = 1, \ldots, n. \tag{2.6}$$

Proof. Let $\mathbf{z}$ be the vector with components z_i. In terms of the representation of a natural cubic spline in terms of its values and second derivatives, the condition (2.6) will be satisfied provided $\mathbf{g} = \mathbf{z}$. By Theorem 2.1, such a natural cubic spline will exist provided we can find a vector γ such that $Q^T\mathbf{g} = R\gamma$. Since R is strictly positive-definite, there will be a unique γ, given by $\gamma = R^{-1}Q^T\mathbf{g}$, satisfying the required condition. □

The theorem can now be used as the basis for a practical algorithm for finding the natural cubic spline interpolant to a set of n data points in $O(n)$ operations. The matrix R is a *tridiagonal matrix*, in other words $r_{ij} = 0$ if $|i - j| \geq 2$, and hence the vector equation $R\gamma = \mathbf{x}$ can be solved for γ in a linear number of operations without finding R^{-1}; for details see any standard numerical linear algebra computer package or textbook. The tridiagonal nature of Q means that $Q^T\mathbf{g}$ can be found from $\mathbf{g}$ in a linear number of operations. Probably the easiest way to calculate $Q^T\mathbf{g}$ is to notice that, for $i = 2, \ldots, n - 1$,

$$(Q^T\mathbf{g})_i = \frac{g_{i+1} - g_i}{h_i} - \frac{g_i - g_{i-1}}{h_{i-1}}, \tag{2.7}$$

so that premultiplication by Q^T is achieved by differencing, dividing componentwise by the values h_j, and differencing again.

We can now conclude that the following algorithm will yield the natural cubic spline interpolant to n points (t_i, z_i) in $O(n)$ computer floating point operations, provided an appropriate numerical method is used in Step 2.

Algorithm for natural cubic spline interpolation

Step 1 Set $g_i = z_i$ for $i = 1, \ldots, n$.
Step 2 Set $\mathbf{x} = Q^T\mathbf{g}$ (by using the formula (2.7)) and solve $R\gamma = \mathbf{x}$ for γ.

2.2.2 Optimality properties of the natural cubic spline interpolant

It has already been pointed out in Section 2.2 that the natural cubic spline interpolant has the important property of having the minimum value of

$\int g''^2$ among all smooth curves that interpolate the data. In the present section we shall give a proof of this assertion using simple calculus; the reader prepared to take the mathematical details on trust should feel free to skip over them.

It should be stressed that the spline nature of the minimizing curve is a mathematical consequence of the choice of roughness penalty functional $\int g''^2$. One of the attractions of the natural cubic spline interpolation method is, of course, the happy combination of circumstances both that the estimate is the solution of a neatly expressed and intuitively attractive minimization, and that it can be calculated in linear time and stored easily as a piecewise polynomial. Neither the natural spline structure itself, nor the positioning of the knots at the points t_i, is imposed on the minimizing curve.

For mathematical completeness, we shall show that the natural cubic spline interpolant is optimal over an even larger class of smooth functions than that considered in Section 2.2. Let $\mathcal{S}_2[a,b]$ be the space of functions that are differentiable on $[a,b]$ and have absolutely continuous first derivative; this means that g is continuous and differentiable everywhere on $[a,b]$ with derivative g', and that there is an integrable function g'' such that $\int_a^x g''(t)dt = g'(x) - g'(a)$ for all x in $[a,b]$. This condition is automatically satisfied if g is continuously twice differentiable on $[a,b]$, and so $\mathcal{S}_2[a,b]$ contains all the functions in $\mathcal{S}[a,b]$.

We can now state and prove the main theorem of this section, which shows that the natural cubic spline interpolant is the unique minimizer of $\int g''^2$ over the class of all functions in $\mathcal{S}_2[a,b]$ that interpolate the data.

Theorem 2.3 *Suppose $n \geq 2$, and that g is the natural cubic spline interpolant to the values $z_1, \ldots, z_n$ at points $t_1, \ldots, t_n$ satisfying $a < t_1 < \ldots < t_n < b$. Let $\tilde{g}$ be any function in $\mathcal{S}_2[a,b]$ for which $\tilde{g}(t_i) = z_i$ for $i = 1, \ldots, n$. Then $\int \tilde{g}''^2 \geq \int g''^2$, with equality only if $\tilde{g}$ and g are identical.*

Proof. Let h be the function in $\mathcal{S}_2[a,b]$ given by $h = \tilde{g} - g$. Both $\tilde{g}$ and g interpolate the values z_i, and so h is zero at all the points t_i for $i = 1, \ldots, n$. Since, by the natural boundary conditions, g'' is zero at a and b, integration by parts yields

$$\begin{aligned}
\int_a^b g''(t)h''(t)dt &= -\int_a^b g'''(t)h'(t)dt \\
&= -\sum_{j=1}^{n-1} g'''(t_j^+) \int_{t_j}^{t_{j+1}} h'(t)dt \\
&= -\sum_{j=1}^{n-1} g'''(t_j^+)\{h(t_{j+1}) - h(t_j)\} = 0. \qquad (2.8)
\end{aligned}$$

We have used the fact that g''' is zero on each of the intervals (a, t_1) and (t_n, b), and is constant on each interval (t_j, t_{j+1}) with value $g'''(t_j^+)$.

It follows, substituting (2.8), that

$$\begin{aligned} \int_a^b \tilde{g}''^2 &= \int_a^b (g'' + h'')^2 = \int_a^b g''^2 + 2\int_a^b g''h'' + \int_a^b h''^2 \\ &= \int_a^b g''^2 + \int_a^b h''^2 \geq \int_a^b g''^2, \end{aligned} \tag{2.9}$$

as required. Equality will hold in (2.9) only if $\int h''^2$ is zero, so that h is linear on $[a, b]$. But since h is zero at the points $t_1, ..., t_n$, and since $n \geq 2$, this can only happen if h is identically zero, in other words if g and $\tilde{g}$ are the same function. This completes the proof of the theorem. □

2.3 Smoothing splines

We now return to the more statistical question of constructing an estimate of a curve whose values are observed subject to random error. As in Section 2.2, suppose that $t_1, ..., t_n$ are points in $[a, b]$ satisfying $a < t_1 < ... < t_n < b$. Suppose that we have observations $Y_1, ..., Y_n$. We shall assume throughout this section that $n \geq 3$, in order to ensure that none of the matrices or conditions in Theorem 2.1 are vacuous, but some remarks about $n = 1$ and 2 are made in Section 2.3.4 below. Given any function g in $\mathcal{S}_2[a, b]$, let $S(g)$ be the penalized sum of squares

$$\sum_{i=1}^{n} \{Y_i - g(t_i)\}^2 + \alpha \int_a^b \{g''(x)\}^2 dx$$

as defined in (1.3), with positive smoothing parameter α.

The curve estimate $\hat{g}$ will be defined to be the minimizer of $S(g)$ over the class $\mathcal{S}_2[a, b]$ of all sufficiently smooth curves on $[a, b]$. In this section we shall explore the properties of $\hat{g}$ and explain one way in which it can be calculated. We shall not, for the moment, discuss the choice of the smoothing parameter α; this question will be considered in detail later on.

In the next two sections we shall develop the properties of $\hat{g}$ and then summarize these in a formal theorem at the end.

2.3.1 *Restricting the class of functions to be considered*

Our work in Section 2.2 on interpolating splines was somewhat laborious but it now enables us to obtain, with very little effort, important properties of the estimate $\hat{g}$. First of all we can show that $\hat{g}$ is necessarily a natural cubic spline with knots at the points t_i, and this will be done in the next paragraph.

Suppose g is any curve that is not a natural cubic spline with knots at the t_i. Let $\bar{g}$ be the natural cubic spline interpolant to the values $g(t_i)$; since, by definition, $\bar{g}(t_i) = g(t_i)$ for all i, it is immediate that $\sum\{Y_i - \bar{g}(t_i)\}^2 = \sum\{Y_i - g(t_i)\}^2$. Because of the optimality properties of the natural cubic spline interpolant, $\int \bar{g}''^2 < \int g''^2$, and hence, since $\alpha > 0$, we can conclude that $S(\bar{g}) < S(g)$. This means that, unless g itself is a natural cubic spline, we can find a natural cubic spline which attains a smaller value of the penalized sum of squares (1.3); it follows at once that the minimizer $\hat{g}$ of S must be a natural cubic spline.

It is important to notice that we have not forced $\hat{g}$ to be a natural cubic spline. Just as in the case of interpolation, the natural cubic spline properties of $\hat{g}$ arise as a mathematical consequence of the choice of $\int g''^2$ as a roughness penalty.

Knowing that $\hat{g}$ is a natural cubic spline is an enormous advance. We can specify $\hat{g}$ exactly by finding a finite number of constants because we now only need to minimize $S(g)$ over a finite-dimensional class of functions, the natural cubic splines with knots at the t_i, instead of considering the infinite dimensional set of smooth functions $\mathcal{S}_2[a, b]$. In the next section we shall show how the minimizing spline curve can be found by solving a set of linear equations.

2.3.2 *Existence and uniqueness of the minimizing spline curve*

Suppose, now, that g is a natural cubic spline defined as in Section 2.1.2, with vectors $\mathbf{g}$ and γ, and matrices Q and R, as defined there. In this section we re-express $S(g)$ in terms of these vectors and matrices. We shall be able to conclude that the minimizer $\hat{g}$ exists and is unique, and furthermore we shall be able to give a linear time algorithm for its calculation.

Let $\mathbf{Y}$ be the vector $(Y_1, \ldots, Y_n)^T$. It is immediate that the residual sum of squares about g can be rewritten

$$\sum\{Y_i - g(t_i)\}^2 = (\mathbf{Y} - \mathbf{g})^T(\mathbf{Y} - \mathbf{g})$$

since the vector $\mathbf{g}$ is precisely the vector of values $g(t_i)$. Express the roughness penalty term $\int g''^2$ as $\mathbf{g}^T K \mathbf{g}$ from (2.5) to obtain

$$\begin{aligned} S(g) &= (\mathbf{Y} - \mathbf{g})^T(\mathbf{Y} - \mathbf{g}) + \alpha \mathbf{g}^T K \mathbf{g} \\ &= \mathbf{g}^T(I + \alpha K)\mathbf{g} - 2\mathbf{Y}^T\mathbf{g} + \mathbf{Y}^T\mathbf{Y}. \end{aligned} \qquad (2.10)$$

Since αK is non-negative definite, the matrix $I + \alpha K$ is strictly positive-definite. It therefore follows that (2.10) has a unique minimum, obtained by setting

$$\mathbf{g} = (I + \alpha K)^{-1}\mathbf{Y}. \tag{2.11}$$

To see this, rewrite (2.10) as

$$\{\mathbf{g} - (I + \alpha K)^{-1}\mathbf{Y}\}^T (I + \alpha K) \{\mathbf{g} - (I + \alpha K)^{-1}\mathbf{Y}\} \tag{2.12}$$

plus a constant that depends only on $\mathbf{Y}$; because $I+\alpha K$ is strictly positive-definite, the expression (2.12) is always strictly positive except when $\mathbf{g}$ satisfies (2.11), when it is zero.

We know from Theorem 2.2 that the vector $\mathbf{g}$ defines the spline g uniquely. Thus, over the space of all natural cubic splines with knots at the points t_i, $S(g)$ has a unique minimum given by (2.11). This completes the characterization of the solution to the minimization of the penalized sum of squares; for convenience of reference we summarize the work of the last two sections in the following theorem.

Theorem 2.4 *Suppose $n \geq 3$ and that $t_1, \ldots, t_n$ are points satisfying $a < t_1 < \ldots < t_n < b$. Given data points $Y_1, \ldots, Y_n$, and a strictly positive smoothing parameter α, let $\hat{g}$ be the natural cubic spline with knots at the points $t_1, \ldots, t_n$ for which $\mathbf{g} = (I + \alpha K)^{-1}\mathbf{Y}$. Then, for any g in $\mathcal{S}_2[a, b]$,*

$$S(\hat{g}) \leq S(g)$$

with equality only if g and $\hat{g}$ are identical.

In practice it is inefficient to use (2.11) directly to find the vector $\mathbf{g}$ and hence the curve $\hat{g}$. In the next section we shall use (2.11) to develop a much more efficient algorithm.

2.3.3 The Reinsch algorithm

In this section an algorithm due to Reinsch (1967) for finding the smoothing spline will be obtained.

The basic idea of the Reinsch algorithm is to set up a non-singular system of linear equations for the second derivatives γ_i of $\hat{g}$ at the knots t_i. These equations have a banded structure and can be solved in $O(n)$ arithmetic operations. Explicit formulae then give the values g_i in terms of the γ_i and the data values Y_i. In our discussion, we shall use various ideas from numerical linear algebra such as the Cholesky decomposition of a band matrix; readers unfamiliar with these concepts are referred to Section 2.6.1 below.

A matrix is said to be a *band matrix* if all of its non-zero entries are concentrated on some small number of diagonals; the number of non-zero diagonals is called the bandwidth of the matrix. Thus if B is a symmetric band matrix with bandwidth $2k + 1$, the element B_{ij} is zero if $|i - j| > k$. Band matrices are economical to store since there is no need to allocate storage for the diagonals known to be zero. Diagonal and tridiagonal matrices are of course band matrices with bandwidths 1 and 3 respectively. The matrices R and Q as defined in Section 2.1.2 both have bandwidth 3.

Define the $(n-2)$-vector γ as in Section 2.1.2. From (2.3) and (2.11) we have

$$(I + \alpha Q R^{-1} Q^T)\mathbf{g} = \mathbf{Y}. \tag{2.13}$$

Rearrange (2.13) to obtain

$$\mathbf{g} = \mathbf{Y} - \alpha Q R^{-1} Q^T \mathbf{g};$$

now substitute $Q^T\mathbf{g} = R\gamma$ and simplify to give an explicit formula for $\mathbf{g}$ in terms of $\mathbf{Y}$ and γ

$$\mathbf{g} = \mathbf{Y} - \alpha Q \gamma. \tag{2.14}$$

Again using the condition $Q^T\mathbf{g} = R\gamma$ we obtain

$$Q^T\mathbf{Y} - \alpha Q^T Q \gamma = R\gamma,$$

which gives the equation for γ

$$(R + \alpha Q^T Q)\gamma = Q^T\mathbf{Y}. \tag{2.15}$$

This equation is the core of the algorithm. By contrast to the equation (2.13) for $\mathbf{g}$, it can be solved in linear time using band matrix techniques.

The matrix $(R + \alpha Q^T Q)$ is easily seen to have bandwidth 5, and also to be symmetric and strictly positive-definite. Therefore it has a Cholesky decomposition of the form

$$R + \alpha Q^T Q = LDL^T$$

where D is a strictly positive diagonal matrix and L is a lower triangular band matrix with $L_{ij} = 0$ for $j < i - 2$ and $j > i$, and $L_{ii} = 1$ for all i. The matrices Q and R can all be found in $O(n)$ algebraic operations, provided only the non-zero diagonals are stored, and hence the matrices L and D require only linear time for their computation.

The Reinsch algorithm can now be set out. Because the matrices involved are all band matrices, each step can be performed in $O(n)$ algebraic operations.

Algorithm for spline smoothing

Step 1 Evaluate the vector $Q^T\mathbf{Y}$, by using the formula (2.7).

Step 2 Find the non-zero diagonals of $R + \alpha Q^T Q$, and hence the Cholesky decomposition factors L and D.

Step 3 Write (2.15) as $LDL^T\gamma = Q^T\mathbf{Y}$ and solve this equation for γ by forward and back substitution.

Step 4 From (2.14), use

$$\mathbf{g} = \mathbf{Y} - \alpha Q\gamma$$

to find $\mathbf{g}$.

It is worth remarking that **Step 1** need be performed only once for each data set; it does not need to be repeated if a new value of the smoothing parameter α is used. Furthermore if new data values $\mathbf{Y}$ are used, but the design points left unchanged, then **Step 2** can be omitted.

2.3.4 *Some concluding remarks*

A close inspection of Theorem 2.4 and of the Reinsch algorithm reveals that none of the vectors and matrices considered in the Reinsch algorithm depend on the choice of the interval $[a, b]$. In fact it is not difficult to see why the minimizing curve $\hat{g}$ essentially does not depend on a and b, beyond the condition that all the data points lie between a and b. Suppose that (a^*, b^*) is an extension of the range (a, b), and that $\hat{g}$ is extended to (a^*, b^*) by linear extrapolation at each end. Since $\hat{g}''$ is zero outside (t_1, t_n), $\int_{a^*}^{b^*} \hat{g}''^2 = \int_a^b \hat{g}''^2$, and so $S(\hat{g})$ will remain the same. For any other function g, extending the range of integration of g''^2 will if anything increase $\int g''^2$, and hence $\hat{g}$ will still miminize S. Indeed if $\hat{g}$ is extended linearly to $(-\infty, \infty)$ at each end, then $\hat{g}$ will minimize S over $\mathcal{S}_2[a, b]$ for *any* a and b with $-\infty \leq a \leq t_1$ and $t_n \leq b \leq \infty$. Of course just because it is mathematically possible to extrapolate $\hat{g}$ beyond the range of the data does not mean that extrapolation is statistically desirable. But it is interesting, and comforting, to know that essentially the same smoothing spline will be obtained no matter what the interval containing the data is taken to be.

We have concentrated on the case where there are at least three distinct data points, and a few words should be said about the cases $n = 1$ and 2, which are of course of little practical interest. If $n = 2$, it is immediately clear that, whatever the value of α, setting $\hat{g}$ to be the straight line through the two points (t_1, Y_1) and (t_2, Y_2) will, uniquely, reduce $S(\hat{g})$ to zero. Thus the minimization problem has a unique solution, but an algorithm is scarcely required to find it! In the (even more absurd) case $n = 1$, the minimizer of S is no longer unique, because any straight line through (t_1, Y_1) will yield a zero value of $S(g)$.

2.4 Plotting a natural cubic spline

Suppose that g is a natural cubic spline with knots $t_1 < t_2 <\ldots< t_n$. Let $\mathbf{g}$ be the vector of values of g at the knot points and let γ be the vector of second derivatives at the knot points. In this section we explain in detail how the vectors $\mathbf{g}$ and γ can be used to find the value of g at any point t, and hence to plot a graph of g to any desired degree of accuracy. (Of course, for some purposes it will be perfectly satisfactory to join the points (t_i, g_i) by straight lines.)

Concentrate, to start with, on the interval between any successive pair of knots t_j and t_{j+1}. On this interval, g is a cubic polynomial, and its value and second derivative are specified at the ends of the interval. In the next section we describe how to use this information to construct the cubic on the entire interval.

2.4.1 *Constructing a cubic given values and second derivatives at the ends of an interval*

Suppose g is a cubic on the interval $[t_L, t_R]$ and that

$$g(t_L) = g_L,\ g(t_R) = g_R,\ g''(t_L^+) = \gamma_L,\ \text{ and } g''(t_R^-) = \gamma_R. \tag{2.16}$$

Define $h = t_R - t_L$. Since g is a cubic, it follows at once that g'' is linear on $[t_L, t_R]$. Therefore we have

$$g''(t) = \frac{(t - t_L)\gamma_R + (t_R - t)\gamma_L}{h} \tag{2.17}$$

and, by differentiating,

$$g'''(t) = \frac{\gamma_R - \gamma_L}{h}. \tag{2.18}$$

To get expressions for the value and derivative of g is a little more complicated. It can be shown that

$$\begin{aligned} g(t) &= \frac{(t - t_L)g_R + (t_R - t)g_L}{h} \\ &\quad - \tfrac{1}{6}(t - t_L)(t_R - t)\left\{\left(1 + \frac{t - t_L}{h}\right)\gamma_R + \left(1 + \frac{t_R - t}{h}\right)\gamma_L\right\}. \end{aligned} \tag{2.19}$$

It is immediately clear that the expression given in (2.19) is a cubic and has the required values at t_L and t_R. By straightforward calculus it can be checked that the second derivatives are also correct; the details are left as an exercise for the reader.

We shall not give an explicit expression for the derivative of f at a general point, but note that (again leaving the details as an exercise) it follows from (2.19) that

$$g'(t_L^+) = \frac{g_R - g_L}{h} - \tfrac{1}{6}h(2\gamma_L + \gamma_R) \tag{2.20}$$

and

$$g'(t_R^-) = \frac{g_R - g_L}{h} + \tfrac{1}{6}h(\gamma_L + 2\gamma_R). \tag{2.21}$$

2.4.2 *Plotting the entire cubic spline*

Now return to the full cubic spline g. The expression (2.19) can be used on each interval $[t_i, t_{i+1}]$ in turn to plot g on the interval $[t_1, t_n]$ delimited by the knots. For $i = 1, \ldots, n-1$ define $h_i = t_{i+1} - t_i$. We then have

$$\begin{aligned} g(t) &= \frac{(t-t_i)g_{i+1} + (t_{i+1}-t)g_i}{h_i} \\ &\quad - \tfrac{1}{6}(t-t_i)(t_{i+1}-t)\left\{\left(1 + \frac{t-t_i}{h_i}\right)\gamma_{i+1} + \left(1 + \frac{t_{i+1}-t}{h_i}\right)\gamma_i\right\} \\ &\quad \text{for } t_i \le t \le t_{i+1},\ i = 1, \ldots, n-1. \end{aligned} \tag{2.22}$$

Similarly, the expressions (2.17) and (2.18) give the second and third derivative on each interval.

If $t \le t_1$ or $t \ge t_n$ then the definition of a NCS implies that $g''(t) = g'''(t) = 0$. To get an expression for $g(t)$ itself, we use the fact that g is linear outside the range of the knots. The derivatives of g at t_1 and t_n are, by substituting into (2.20) and (2.21), given by

$$g'(t_1) = \frac{g_2 - g_1}{t_2 - t_1} - \tfrac{1}{6}(t_2 - t_1)\gamma_2$$

and

$$g'(t_n) = \frac{g_n - g_{n-1}}{t_n - t_{n-1}} + \tfrac{1}{6}(t_n - t_{n-1})\gamma_{n-1}.$$

Once these values have been calculated, the linearity of g outside the range of the knots gives

$$g(t) = g_1 - (t_1 - t)g'(t_1) \text{ for } t \le t_1 \tag{2.23}$$

and

$$g(t) = g_n + (t - t_n)g'(t_n) \text{ for } t \ge t_n. \tag{2.24}$$

It should be noted that these expressions do not depend on the overall interval $[a, b]$ of interest, provided this interval contains all the knots.

Any reader interested in computer graphics is advised, as an exercise, to write a program that plots the cubic spline g given the values of the knots and the vectors $\mathbf{g}$ and $\boldsymbol{\gamma}$.

2.5 Some background technical properties

In Theorem 2.1 it was stated that the vectors $\mathbf{g}$ and $\boldsymbol{\gamma}$ specify a natural cubic spline if and only if $Q^T\mathbf{g} = R\boldsymbol{\gamma}$, and expressions were given for the integrated squared second derivative of the resulting NCS if the condition is satisfied. In this section we prove this theorem.

2.5.1 *The key property for g to be a natural cubic spline*

Given $\mathbf{g}$ and $\boldsymbol{\gamma}$, define $g(t)$ to be the piecewise cubic polynomial defined by (2.22) for $t_1 \leq t \leq t_n$ and by (2.23) and (2.24) outside this range. It is immediate from the construction that g is everywhere continuous, that $g''(t_j^+) = g''(t_j^-) = \gamma_j$ for each j and that g' is continuous at t_1 and t_n. The only way in which g can fail to be a natural cubic spline is if the cubic pieces do not fit together at the internal knots $t_2, \ldots, t_{n-1}$ in a differentiable way.

For each j, $2 \leq j \leq n-1$, it follows from (2.21) and (2.20) respectively that

$$g'(t_j^-) = \frac{g_j - g_{j-1}}{h_{j-1}} + \tfrac{1}{6}h_{j-1}(\gamma_{j-1} + 2\gamma_j) \tag{2.25}$$

and

$$g'(t_j^+) = \frac{g_{j+1} - g_j}{h_j} - \tfrac{1}{6}h_j(2\gamma_j + \gamma_{j+1}). \tag{2.26}$$

Thus the derivative of the piecewise polynomial g will be continuous at all the knots (and hence g will be a natural cubic spline) if and only if, for $j = 2, \ldots, n-1$, the expressions (2.25) and (2.26) are equal; rearranging, this yields

$$\frac{g_{j+1} - g_j}{h_j} - \frac{g_j - g_{j-1}}{h_{j-1}} = \tfrac{1}{6}h_{j-1}\gamma_{j-1} + \tfrac{1}{3}(h_{j-1} + h_j)\gamma_j + \tfrac{1}{6}h_j\gamma_{j+1}, \tag{2.27}$$

precisely the condition $Q^T\mathbf{g} = R\boldsymbol{\gamma}$ as in (2.4), as required.

2.5.2 *Expressions for the roughness penalty*

Now suppose that the condition (2.4) is satisfied. Just as in (2.8), integrate by parts, use the facts that $g''(a) = g''(b) = 0$, and that g''' is constant on each interval (t_j, t_{j+1}) and zero outside $[t_1, t_n]$, to obtain

$$\begin{aligned}\int_a^b \{g''(t)\}^2 dt &= -\int_a^b g'''(t)g'(t)dt \\ &= -\sum_{j=1}^{n-1} g'''(t_j^+) \int_{t_j}^{t_{j+1}} g'(t)dt\end{aligned}$$

$$= \sum_{j=1}^{n-1} \frac{(\gamma_{j+1} - \gamma_j)}{h_j}(g_j - g_{j+1}) \qquad (2.28)$$

by substituting (2.18) for the third derivative of g. Since $\gamma_1 = \gamma_n = 0$, it follows by rearranging (2.28) that

$$\int_a^b \{g''(t)\}^2 dt = \sum_{i=2}^{n-1} \gamma_i \left(\frac{g_{i+1} - g_i}{h_i} - \frac{g_i - g_{i-1}}{h_{i-1}} \right) = \gamma^T Q^T \mathbf{g}$$

$$= \gamma^T R \gamma = \mathbf{g}^T Q R^{-1} Q^T \mathbf{g} = \mathbf{g}^T K \mathbf{g},$$

making use of the property $Q^T \mathbf{g} = R\gamma$ and the definition (2.3) of K. This completes the proof of Theorem 2.1.

2.6 Band matrix manipulations

The fast algorithms for calculating the cubic spline interpolant and the spline smoother make use of band matrix manipulations.

For the benefit of readers unfamiliar with band matrices, we briefly review some of their properties in this section. We do not attempt a complete treatment, but concentrate on the techniques used in algorithms for spline smoothing and interpolation. Routines for manipulating band matrices are given in most mathematical subroutine packages, and so anybody contemplating the implementation of the techniques described here should first of all check whether the required routine is already available. The algorithms we describe are known to be numerically stable; for details and references to further reading see, for example, Golub and Van Loan (1983).

In our treatment we shall focus attention on the case of bandwidth 5, since this is the value relevant to the matrices that arise in constructing the smoothing spline, for example using the Reinsch algorithm; see Section 2.3.3 above.

2.6.1 The Cholesky decomposition

Suppose, now, that B is a positive-definite symmetric $n \times n$ band matrix. It is then possible to express B as

$$B = LDL^T \qquad (2.29)$$

where D is a diagonal matrix with diagonal elements D_i, and L is a lower triangular band matrix with diagonal elements all equal to 1. The decomposition (2.29) is called the Cholesky decomposition, and is useful for a number of purposes, particularly for solving systems of linear equations as described in Section 2.6.2 below.

In order to perform the decomposition, we equate the non-zero elements of B and LDL^T on and below the main diagonal, row by row in the order $(1,1), (2,1), (2,2), (3,1), (3,2), (3,3), \ldots, (n,n-2), (n,n-1), (n,n)$. In the particular case of bandwidth 5, this gives, after some rearrangement,

$$D_1 = B_{11},\ L_{21} = B_{21}/D_1,\ D_2 = B_{22} - L_{21}^2 D_1$$

and, for $i = 3, \ldots, n$ successively,

$$\begin{aligned} L_{i,i-2} &= B_{i,i-2}/D_{i-2}, \\ L_{i,i-1} &= (B_{i,i-1} - L_{i-1,i-2}L_{i,i-2}D_{i-2})/D_{i-1}, \text{ and} \\ D_i &= B_{ii} - L_{i,i-1}^2 D_{i-1} - L_{i,i-2}^2 D_{i-2}. \end{aligned}$$

Since each element of B is referred to only once in the procedure we have described, the decomposition can be performed 'in place' with D_i placed in the location occupied by B_{ii} and $L_{i,i-2}$ and $L_{i,i-1}$ placed in the locations of $B_{i,i-2}$ and $B_{i,i-1}$. It can be seen by counting up the operations required that the entire decomposition can be carried out in $O(n)$ arithmetic operations; $8n - 13$ multiplications/divisions are needed for the calculations as set out above in the case of bandwidth 5.

2.6.2 Solving linear equations

Suppose, as in Section 2.6.1, that B is a positive-definite symmetric $n \times n$ band matrix. Suppose that $\mathbf{z}$ is a known n-vector, and that it is of interest to solve the system of equations

$$B\mathbf{x} = \mathbf{z} \tag{2.30}$$

for the unknown n-vector $\mathbf{x}$. The Cholesky decomposition makes it easy to solve these equations very quickly. As in (2.29), decompose B as LDL^T and introduce vectors $\mathbf{u}$ and $\mathbf{v}$ satisfying

$$L\mathbf{u} = \mathbf{z},\ D\mathbf{v} = \mathbf{u} \text{ and } L^T\mathbf{x} = \mathbf{v}.$$

Next, find the components of $\mathbf{u}$ in order:

$$u_1 = z_1,\ u_2 = z_2 - L_{21}u_1, \text{ and}$$

$$u_i = z_i - L_{i,i-1}u_{i-1} - L_{i,i-2}u_{i-2} \text{ for } i = 3, \ldots, n;$$

then set $v_i = u_i/D_i$ for each i; and finally equate coefficients in reverse order to find $\mathbf{x}$:

$$x_n = v_n,\ x_{n-1} = v_{n-1} - L_{n,n-1}x_n, \text{ and}$$

$$x_i = v_i - L_{i+1,i}x_{i+1} - L_{i+2,i}x_{i+2} \text{ for } i = n-2, n-3, \ldots, 1.$$

Any or all of these three steps can be performed in place with the components, for example, of $\mathbf{u}, \mathbf{v}$ and $\mathbf{x}$ all occupying the same locations successively. Once the Cholesky decomposition has been performed, it can be seen at once that the number of operations required to solve the equations (2.30) is again $O(n)$; for the case of bandwidth 5 that we have described in detail, the number of multiplications/divisions needed is $5n - 6$.

The cubic spline interpolation algorithm described in Section 2.2 includes the solution of a set of equations of the form (2.30) with B a tridiagonal matrix, that is a matrix whose bandwidth is 3. We leave as an exercise the simplification to this case of the procedures we have set out above; altogether the number of multiplications/divisions required is $3n - 3$ for the decomposition and an additional $3n - 2$ for the solution itself.

CHAPTER 3

One-dimensional case: further topics

3.1 Choosing the smoothing parameter

The problem of choosing the smoothing parameter is ubiquitous in curve estimation, even if in certain cases it is swept under the carpet in the way a method is specified. For example, if one is fitting curves by polynomial regression, the choice of the degree of the fitted polynomial is essentially equivalent to the choice of a smoothing parameter. In the spline smoothing methodology set out in Chapter 2 above, the smoothing parameter is of course explicit in the method.

There are two different philosophical approaches to the question of choosing the smoothing parameter. The first approach is to regard the free choice of smoothing parameter as an advantageous feature of the procedure. By varying the smoothing parameter features of the data that arise on different 'scales' can be explored, and if a single estimate is ultimately needed it can be obtained by a subjective choice. It may well be that such a subjective approach is in reality the most useful one.

The other, to some extent opposing, philosophical view is that there is a need for an automatic method whereby the smoothing parameter value is chosen by the data. It is fairer to use the word *automatic* rather than *objective* for such a method, because—as in almost any statistical procedure—there are arbitrary decisions involved in the choice of the method itself. Nevertheless it is of course the case that conditionally on the automatic method being used the choice of smoothing parameter is indeed objective.

Automatic methods need not be used in an uncritical way; they can of course be used as a starting point for fine tuning. They are almost essential if the estimated curve is to be used as a component part of a more complicated procedure, or if the method is being used routinely on a large number of data sets. In the latter context it may well be that there is a preference for using the same value of the smoothing parameter across different data sets in order to aid comparison, or simply because a particular value is known from experience to give good results!

There are a number of different automatic procedures available. Probably the most well known is cross-validation, which will be described in the next section.

3.2 Cross-validation

The basic motivation behind the cross-validation method is in terms of prediction. Assuming that the random error has zero mean, the true regression curve g has the property that, if an observation Y is taken at a point t, the value $g(t)$ is the best predictor of Y in terms of mean square error. Thus a good choice of estimator $\hat{g}(t)$ would be one that gave a small value of $\{Y - \hat{g}(t)\}^2$ for a new observation Y at the point t.

Of course in practice, when the smoothing method is applied to a single data set, no new observations are available. The cross-validation technique manufactures the 'new observation' situation from the given data as follows.

Focus attention on a given smoothing parameter value α. Let us consider the observation Y_i at t_i as being a new observation by omitting it from the set of data used to estimate the curve itself. Denote by $\hat{g}^{(-i)}(t;\alpha)$ the curve estimated from the remaining data, using the value α for the smoothing parameter, so that $\hat{g}^{(-i)}(t;\alpha)$ is the minimizer of

$$\sum_{j \neq i} \{Y_j - g(t_j)\}^2 + \alpha \int g''^2. \tag{3.1}$$

The quality of $\hat{g}^{(-i)}$ as a predictor on a new observation can be judged by how well the value $\hat{g}^{(-i)}(t_i)$ predicts Y_i. Since the choice of which observation to omit is arbitrary, the overall efficacy of the procedure with the smoothing parameter α can be quantified by the cross-validation score function

$$CV(\alpha) = n^{-1} \sum_{i=1}^{n} \{Y_i - \hat{g}^{(-i)}(t_i;\alpha)\}^2. \tag{3.2}$$

The basic idea of cross-validation is to choose the value of α that minimizes $CV(\alpha)$. It cannot be guaranteed that the function CV has a unique minimum, so care has to be taken with its minimization, and a simple grid search is probably the best approach. Whatever minimization method is used, it will involve calculating $CV(\alpha)$ for a number of values of α, and therefore an efficient method of calculating CV is important.

3.2.1 Efficient calculation of the cross-validation score

At first sight, it would appear from (3.2) that to work out $CV(\alpha)$ it is necessary to solve n separate smoothing problems, in order to find the n curves $\hat{g}^{(-i)}$. However, we shall see in this section that this is by no means the case.

The first step in the simplification is to recall from Theorem 2.4 that the values of the smoothing spline $\hat{g}$ depend linearly on the data Y_i through the equation

$$\mathbf{g} = A(\alpha)\mathbf{Y} \tag{3.3}$$

where the matrix $A(\alpha)$ is defined by

$$A(\alpha) = (I + \alpha Q R^{-1} Q^T)^{-1}. \tag{3.4}$$

The matrix $A(\alpha)$ is called the *hat matrix* because it maps the vector of observed values Y_i to their 'predicted values' $\hat{g}(t_i)$ or $\hat{Y}_i$.

The first key result in the development of an economical way of calculating the cross-validation score is as follows.

Theorem 3.1 *The cross-validation score satisfies*

$$CV(\alpha) = n^{-1} \sum_{i=1}^{n} \left(\frac{Y_i - \hat{g}(t_i)}{1 - A_{ii}(\alpha)} \right)^2, \tag{3.5}$$

where $\hat{g}$ is the spline smoother calculated from the full data set $\{(t_i, Y_i)\}$ with smoothing parameter α.

This theorem shows that, provided the diagonal entries $A_{ii}(\alpha)$ are known, the cross-validation score can be calculated from the residuals $Y_i - \hat{g}(t_i)$ about the spline smoother calculated from the full data set. Therefore no additional smoothing problems have to be solved. However, it would appear from (3. 4) that the computation of the diagonal entries of the hat matrix could be a burdensome task, because of the matrix inversions involved in its definition. In fact this is not the case, and in Section 3.2.2 below we shall describe an ingenious algorithm that yields all the diagonal elements of $A(\alpha)$ in $O(n)$ operations, and hence, since the Reinsch algorithm gives the values $\hat{g}(t_i)$ in linear time, the cross-validation score itself can be found in linear time for each value of α.

Proof of Theorem 3.1

The proof of Theorem 3.1 follows immediately from a lemma that is very well known in other contexts, and gives as a corollary an expression for the 'deleted residual' $Y_i - \hat{g}^{(-i)}(t_i)$ in terms of $Y_i - \hat{g}(t_i)$ and the ith diagonal element of the hat matrix. The result itself is identical in form to that obtained in the development of the PRESS criterion for deciding the complexity of a parametric multivariate regression; see Cook and Weisberg (1982) for example. However we shall give a proof of the lemma, because the details in the smoothing context are slightly non-standard.

Lemma 3.1 *For fixed α and i, let $\mathbf{g}^{(-i)}$ denote the vector with components $g_j^{(-i)} = \hat{g}^{(-i)}(t_j; \alpha)$, and define a vector $\mathbf{Y}^*$ by*

$$\begin{aligned} Y_j^* &= Y_j \text{ for } j \neq i \\ Y_i^* &= \hat{g}^{(-i)}(t_i). \end{aligned}$$

Then

$$\mathbf{g}^{(-i)} = A(\alpha)\mathbf{Y}^*. \tag{3.6}$$

Proof. For any smooth curve g,

$$\begin{aligned} \sum_{j=1}^{n} \{Y_j^* - g(t_j)\}^2 + \alpha \int g''^2 &\geq \sum_{j \neq i} \{Y_j^* - g(t_j)\}^2 + \alpha \int g''^2 \\ &\geq \sum_{j \neq i} \{Y_j^* - \hat{g}^{(-i)}(t_j)\}^2 + \alpha \int \hat{g}^{(-i)''2} \\ &= \sum_{j=1}^{n} \{Y_j^* - \hat{g}^{(-i)}(t_j)\}^2 + \alpha \int \hat{g}^{(-i)''2} \end{aligned}$$

by the definition of $\hat{g}^{(-i)}$ and the fact that $Y_i^* = \hat{g}^{(-i)}(t_i)$. It follows that $\hat{g}^{(-i)}$ is the minimizer of $\sum_{j=1}^{n} \{Y_j^* - g(t_j)\}^2 + \alpha \int g''^2$ so that $\mathbf{g}^{(-i)} = A(\alpha)\mathbf{Y}^*$, as required. □

As a corollary, we can obtain an expression for the deleted residual $Y_i - \hat{g}^{(-i)}(t_i)$. We have (writing A for $A(\alpha)$ throughout)

$$\begin{aligned} \hat{g}^{(-i)}(t_i) - Y_i &= \sum_{j=1}^{n} A_{ij} Y_j^* - Y_i = \sum_{j \neq i} A_{ij} Y_j + A_{ii} \hat{g}^{(-i)}(t_i) - Y_i \\ &= \sum_{j=1}^{n} A_{ij} Y_j - Y_i + A_{ii} \{\hat{g}^{(-i)}(t_i) - Y_i\} \\ &= \hat{g}(t_i) - Y_i + A_{ii} \{\hat{g}^{(-i)}(t_i) - Y_i\}. \end{aligned} \tag{3.7}$$

It follows at once from (3.7) that

$$Y_i - \hat{g}^{(-i)}(t_i) = \frac{Y_i - \hat{g}(t_i)}{1 - A_{ii}(\alpha)}. \tag{3.8}$$

Squaring (3.8) and summing over i gives

$$CV(\alpha) = n^{-1} \sum_{i=1}^{n} \left(\frac{Y_i - \hat{g}(t_i)}{1 - A_{ii}(\alpha)} \right)^2 ,$$

completing the proof of Theorem 3.1.

3.2.2 Finding the diagonal elements of the hat matrix

In this section we describe an algorithm due to Hutchison and de Hoog (1985) for finding the diagonal elements of the hat matrix $A(\alpha)$. The method finds all the elements of the diagonal in $O(n)$ operations. There are two important components of the method, the first a general method for finding the central diagonals of the inverse of a band matrix, and the second a re-expression of the hat matrix in a form that can be calculated without inverting any full matrices. Both of these components apply to matrices of arbitrary bandwidth, but will be described here only for the special case of relevance to cubic spline smoothing, where the bandwidth is 5. The extension to the general case takes the obvious form.

The central diagonals of the inverse of a band matrix

Suppose that B is a symmetric positive-definite band matrix with bandwidth 5, so that the (i,j) element of B is zero if $|i-j| > 2$. The problem of finding B^{-1} requires $O(n^2)$ operations, but there is a linear time algorithm for finding just the central five diagonals of B^{-1}, which we describe in this section.

Decompose B as $B = LDL^T$ where L is a lower triangular band matrix with unit diagonal, and D is a diagonal matrix with elements d_i. Suppose that B^{-1} has elements $\bar{b}_{ij}$. Then, by definition,

$$B^{-1} = L^{-T} D^{-1} L^{-1},$$

so that

$$L^T B^{-1} = D^{-1} L^{-1},$$

from which it follows that

$$B^{-1} = D^{-1}L^{-1} + B^{-1} - L^T B^{-1} = D^{-1}L^{-1} + (I - L^T)B^{-1}. \qquad (3.9)$$

The matrix L^{-1} is a lower triangular matrix with unit diagonal, and therefore $D^{-1}L^{-1}$ is lower triangular with diagonal elements d_i^{-1}. Furthermore $(I - L^T)$ is upper triangular with zero diagonal. Hence it follows from (3.9), equating elements in the main and two leading upper diagonals, and using the symmetry of B^{-1} to write $\bar{b}_{ij} = \bar{b}_{ji}$ whenever $i > j$, that, for $i = 1, \ldots, n-2$,

$$\begin{aligned} \bar{b}_{ii} &= d_i^{-1} - L_{i+1,i}\bar{b}_{i,i+1} - L_{i+2,i}\bar{b}_{i,i+2}, \\ \bar{b}_{i,i+1} &= -L_{i+1,i}\bar{b}_{i+1,i+1} - L_{i+2,i+1}\bar{b}_{i+1,i+2}, \quad \text{and} \\ \bar{b}_{i,i+2} &= -L_{i+1,i}\bar{b}_{i+2,i+2} \end{aligned}$$

and that

$$\begin{aligned} \bar{b}_{n,n} &= d_n^{-1}; \\ \bar{b}_{n-1,n} &= -L_{n,n-1}\bar{b}_{n,n}; \\ \bar{b}_{n-1,n-1} &= d_{n-1}^{-1} - L_{n,n-1}\bar{b}_{n-1,n}. \end{aligned}$$

These formulae can be applied in an appropriate order to find the five central diagonals of B^{-1}. The order in which the elements can be found is $(n,n),(n-1,n),(n-2,n);(n-1,n-1),(n-2,n-1),(n-3,n-1);\ldots;$ $(2,2),(2,1);(1,1)$. Once the elements d_i^{-1} and L_{ij} have been found, all the elements $\bar{b}_{ij}$ for $|i-j| \leq 2$ can be found in $6n-10$ arithmetic operations. It should be stressed again that the elements of B^{-1} further from the diagonal are *not* zero, in general; but we shall see below that they are not relevant to our purposes.

Expressing the hat matrix in suitable form

From equations (2.14) and (2.15) in the description of the Reinsch algorithm in Section 2.3.3 above, the spline smoother is defined by vectors **g** and γ that satisfy

$$\gamma = (R + \alpha Q^T Q)^{-1} Q^T \mathbf{Y}$$

and

$$\begin{aligned} \mathbf{g} &= \mathbf{Y} - \alpha Q\gamma = \mathbf{Y} - \alpha Q(R + \alpha Q^T Q)^{-1} Q^T \mathbf{Y} \\ &= \{I - \alpha Q(R + \alpha Q^T Q)^{-1} Q^T\}\mathbf{Y}. \end{aligned} \tag{3.10}$$

It can be seen from (3.10) that the hat matrix has the alternative expression

$$A(\alpha) = I - \alpha Q(R + \alpha Q^T Q)^{-1} Q^T$$

so that

$$I - A(\alpha) = \alpha Q(R + \alpha Q^T Q)^{-1} Q^T. \tag{3.11}$$

It is of course the diagonal elements of $I - A(\alpha)$ that are needed in the expression (3.5) for the cross-validation score. It can easily be checked by direct matrix manipulations that (3.11) and (3.4) are consistent with one another.

Now define B to be the symmetric band matrix $(R + \alpha Q^T Q)$, which has bandwidth 5. Denote by $\bar{b}_{ij}$ the elements of B^{-1}. Since Q is a tridiagonal matrix, it follows that, for each $i = 1, \ldots, n$ (defining all 'out of range' elements to be zero)

$$\begin{aligned}(QB^{-1}Q^T)_{ii} &= q_{i,i-1}^2\bar{b}_{i-1,i-1} + q_{ii}^2\bar{b}_{ii} + q_{i,i+1}^2\bar{b}_{i+1,i+1} + 2q_{i,i-1}q_{ii}\bar{b}_{i-1,i} \\ &\quad + 2q_{i,i-1}q_{i,i+1}\bar{b}_{i-1,i+1} + 2q_{ii}q_{i,i+1}\bar{b}_{i,i+1}. \qquad (3.12)\end{aligned}$$

It is clear from (3.12) that only those elements $\bar{b}_{ij}$ with $|i-j| \leq 2$ need to be known in order to find the diagonal elements of $QB^{-1}Q^T$, and hence the values $1 - A_{ii}(\alpha)$ needed to calculate the cross-validation score.

3.3 Generalized cross-validation

3.3.1 The basic idea

Generalized cross-validation (GCV), a modified form of cross-validation, is a popular method for choosing the smoothing parameter. An important early reference is Craven and Wahba (1979).

Equation (3.8) above shows that the deleted residuals required for the calculation of the cross-validation score can be obtained from the ordinary residuals by dividing by the factors $1 - A_{ii}(\alpha)$. The basic idea of GCV is to replace these factors by their average value, $1 - n^{-1} \operatorname{tr} A(\alpha)$. The generalized cross-validation score is then constructed, by analogy with ordinary cross-validation, by summing the squared residuals corrected by the square of this factor. Since the factor is the same for all i, we obtain

$$GCV(\alpha) = n^{-1}\frac{\sum_{i=1}^n \{Y_i - \hat{g}(t_i)\}^2}{\{1 - n^{-1} \operatorname{tr} A(\alpha)\}^2}, \qquad (3.13)$$

the residual sum of squares about $\hat{g}$ divided by a correction factor of $n\{1-n^{-1} \operatorname{tr} A(\alpha)\}^2$. Just as in ordinary cross-validation, the GCV choice of smoothing parameter is then carried out by minimizing the function $GCV(\alpha)$ over α.

If all the $A_{ii}(\alpha)$ were equal, for example if the t_i were equally spaced on an interval subject to periodic boundary conditions, then of course the GCV score would be identical to $CV(\alpha)$. More generally, there will be some difference between the two approaches.

3.3.2 Computational aspects

One of the reasons for the original introduction of GCV was computational. By using the alternative expression for the trace of a matrix as the sum of its eigenvalues, it is possible to find the trace of $A(\alpha)$ without finding all its diagonal elements. Suppose that the matrix $QR^{-1}Q^T$ has eigenvalues ω_v; then it follows from (3.4) that $A(\alpha)$ will have eigenvalues

$(1+\alpha\omega_\nu)^{-1}$, and hence that

$$n\{1-n^{-1}\operatorname{tr} A(\alpha)\}^2 = n\left(1-n^{-1}\sum_{\nu=1}^{n}\frac{1}{1+\alpha\omega_\nu}\right)^2. \tag{3.14}$$

Thus once the eigenvalues ω_ν have been found, the GCV score can be found from the residual sum of squares for *any* α by carrying out the simple calculation (3.14). For efficient computational approaches to the exact and approximate calculation of the eigenvalues see Utreras (1980) and Silverman (1984b). Suppose that the design points have 'local density' $f(t)$ on the interval $[a,b]$, in that the proportion of t_i in an interval of length dt near t is approximately $f(t)dt$. (There is no need to assume that the t_i are in any way random, though of course t_i that are randomly distributed with density $f(t)$ would satisfy the condition.) If the t_i are equally spaced then f is just the constant $1/(b-a)$ while more generally f can be estimated in some way. Define the constant c by

$$c = n^{\frac{1}{4}}\pi^{-1}\int_a^b f(t)^{\frac{1}{4}}\,dt.$$

Then it can be shown that $\omega_1 = \omega_2 = 0$ and that $\omega_\nu \approx c^{-4}(\nu-1.5)^4$, so that

$$\operatorname{tr} A(\alpha) \approx 2 + \sum_{\nu=3}^{n}\{1+c^{-4}\alpha(\nu-1.5)^4\}^{-1},$$

a very simple calculation to carry out.

To some extent these economies of calculation have been superseded by increasing computing power and by the Hutchison–de Hoog algorithm set out in Section 3.2.2 above. Since the individual diagonal elements of the hat matrix can be found very rapidly it is easy to find $\operatorname{tr} A(\alpha)$ directly, and there is little computational advantage to be gained by the alternative calculations set out in this section. Choice between GCV and ordinary cross-validation should be based on statistical rather than computational grounds.

3.3.3 Leverage values

In the standard regression literature, for example Cook and Weisberg (1982), the diagonal elements A_{ii} of the hat matrix are called *leverage values*. They determine the amount by which the predicted value $\hat{g}(t_i)$ is influenced by the data value Y_i at the point t_i. At points with a high leverage value the predicted value needs to be treated with some care, because it is particularly sensitive to the observation made at that point.

When the cross-validation score was constructed, all the squared deleted residuals were added together with equal weight. The GCV score can be written in the form

$$GCV(\alpha) = n^{-1} \sum_{i=1}^{n} \left\{ \left(\frac{1 - A_{ii}(\alpha)}{1 - n^{-1} \operatorname{tr} A(\alpha)} \right)^2 \{Y_i - \hat{g}^{(-i)}(t_i)\}^2 \right\},$$

so that GCV can be seen as a modification of cross-validation in which the deleted residuals at points with large leverage values are downweighted somewhat.

3.3.4 *Degrees of freedom*

The connections with classical regression motivate a calculation of 'equivalent degrees of freedom' that give an indication of the effective number of parameters—in some sense—that are fitted for any particular value of the smoothing parameter. Suppose, for a moment, that we were fitting a curve or function g to the data by parametric regression, by assuming $g(t)$ to be of the form $\sum_{j=1}^{k} \theta_j g_j(t)$ for some fixed functions g_j and fitting the parameters θ_j by least squares. (For example, ordinary linear regression is of this form, with $k = 2$, $g_1(t) = 1$ and $g_2(t) = t$.) Assuming the parameters are identifiable on the basis of the available observations, the hat matrix A is then a projection onto a space of dimension k, the number of parameters fitted, and so its trace is equal to k. Thus the model degrees of freedom, k, are equal to trace of A, while the degrees of freedom for noise, $n - k$, are equal to $\operatorname{tr}(I - A)$.

Return now to nonparametric regression. By analogy with the parametric case, we define the *equivalent degrees of freedom for noise (EDF)* by

$$EDF = \operatorname{tr}\{I - A(\alpha)\} \tag{3.15}$$

where, now, $A(\alpha)$ is the hat matrix associated with spline smoothing with smoothing parameter α. The equivalent degrees of freedom for noise increase from 0 when $\alpha = 0$ (interpolation, hat matrix A the identity) to $n - 2$ when $\alpha = \infty$ (linear regression). It follows immediately from the definitions that the GCV score can be written in the form

$$GCV(\alpha) = n \times \frac{\text{residual sum of squares}}{(\text{equivalent degrees of freedom})^2}.$$

The question of definition of equivalent degrees of freedom has been discussed at greater length by Buja, Hastie and Tibshirani (1989); see also Hastie and Tibshirani (1990, Appendix B). In the context of model comparison, they argue for use, not of EDF as defined above, but of $EDF^* = \operatorname{tr}\{(I - A^T)(I - A)\}$. Nevertheless, it is EDF that is appropriate in defining the GCV score. If A is a projection, as in ordinary linear models, the two definitions agree; in other cases, Hastie and Tibshirani suggest a linear approximation $(n - EDF^*) \approx 1.25(n - EDF) - 0.5$ that can be used to avoid an expensive direct calculation of EDF^*.

3.4 Estimating the residual variance

In practice it is very unusual for the residual variance to be known, but for a number of reasons it may be of interest to estimate it. Suppose that our model for the observed data is

$$Y_i = g(t_i) + \epsilon_i \tag{3.16}$$

where the ϵ_i are uncorrelated with zero mean and variance σ^2. There are essentially two different approaches to the estimation of σ^2 that might be considered.

3.4.1 Local differencing

The first possible approach is to transform the observations in such a way that the trend function g is eliminated, to all intents and purposes. For example Rice (1984) suggested using first differences of the data, and hence estimating the variance by

$$\hat{\sigma}_R^2 = \tfrac{1}{2}(n-1)^{-1}\sum_{i=2}^{n}(Y_i - Y_{i-1})^2. \tag{3.17}$$

The rationale behind this estimator is, of course, that for a smooth curve g the first difference $Y_i - Y_{i-1}$ has squared mean $\{g(t_i) - g(t_{i-1})\}^2$ that is small relative to its variance $2\sigma^2$. This will be a reasonable approximation provided the gradient of g is never very large. The use of differencing to eliminate trend is of course very well known in the time series and spatial analysis contexts.

Rice (1984) also suggested a second estimator, based on weighted second differences of the data, that will be invariant under the addition of a linear function to g. This is constructed by fitting a least squares line to successive triples of points. The residual sum of squares about each of these locally fitted lines gives a one degree-of-freedom estimate of σ^2, and averaging these estimates gives the overall estimate of the variance. This is exactly equivalent to an estimate suggested by Gasser, Sroka and Jenner (1986),

$$\hat{\sigma}_{GSJ}^2 = (n-2)^{-1}\sum_{i=2}^{n-1} c_i^2\hat{\epsilon}_i^2, \tag{3.18}$$

where $\hat{\epsilon}_i$ is the difference between Y_i and the value at t_i of the line joining (t_{i-1}, Y_{i-1}) and (t_{i+1}, Y_{i+1}); the constants c_i^2 are chosen to ensure that $E(c_i^2\hat{\epsilon}_i^2) = \sigma^2$ for all i when g is linear. The conceptual advantage of these approaches over (3.17) is that it is natural in the spline smoothing context to assume that the second derivative of g will never be so large as to introduce significant bias into the local estimates $c_i^2\hat{\epsilon}_i^2$ of σ^2.

3.4.2 Residual sum of squares about a fitted curve

The other main class of approaches is to base an estimate of σ^2 on the residual sum of squares about some fitted curve. In parametric regression the standard practice, yielding an unbiased estimator, is to divide the residual sum of squares by the degrees of freedom for noise. The natural analogue in the spline smoothing context is to divide the residual sum of squares by the equivalent degrees of freedom, as defined in Section 3.3.4 above. If the value of the smoothing parameter is α, this yields the estimator

$$\hat{\sigma}_\alpha^2 = \frac{\sum_i \{Y_i - \hat{g}(t_i)\}^2}{\text{tr}\,\{I - A(\alpha)\}} \tag{3.19}$$

where $\hat{g}$ is the spline smoothing estimate calculated with smoothing parameter α. It is easy to show—see, for example, Buckley, Eagleson and Silverman (1988)— that, in the particular case where the true regression function g is a straight line, the estimator $\hat{\sigma}_\alpha^2$ is an unbiased estimator of σ^2 for all α. This provides further motivation for using the term 'equivalent degrees of freedom' for $\text{tr}\,\{I - A(\alpha)\}$.

Intuitively, it would seem appropriate to use the same smoothing parameter when constructing a variance estimator as when estimating the curve itself, so that the estimate of the variance would be obtained as a by-product of the curve estimate. However there might be an argument for smoothing a different amount when estimating the variance. See Besag and Kempton (1986) for discussion of this in the context of agricultural field trials. This question is considered from a slightly different point of view by Buckley, Eagleson and Silverman (1988) and by Carter, Eagleson and Silverman (1992). Their calculations suggest that there may be some merit in basing the variance estimate on a slightly undersmoothed curve estimate, but that little is lost by using the smoothing parameter that is optimal for the estimation of the curve itself.

3.4.3 Some comparisons

Buckley, Eagleson and Silverman (1988) and Carter, Eagleson and Silverman (1992) also make some comparisons between estimators of the form (3.19) and those discussed in Section 3.4.1 above. Consider the asymptotic situation where a large number n of independent normally distributed observations are taken uniformly spaced on a fixed interval, the true regression curve g and the sample variance remaining fixed. It is shown that, for any reasonable choice of smoothing parameter α, the asymptotic mean square error of the estimator $\hat{\sigma}_\alpha^2$ is $2\sigma^4 n^{-1}$, the effect of the choice of smoothing parameter being only on lower order terms. On the other hand, the estimator $\hat{\sigma}_{GSJ}^2$ of (3.18) has asymptotic mean square error $\frac{35}{9}\sigma^4 n^{-1}$, almost twice as large. Rice's estimator $\hat{\sigma}_R^2$ as defined in (3.17) has asymptotic mean square error $3\sigma^4 n^{-1}$.

The intuitive reason for the higher mean square error of the estimators based on local differencing is that they eliminate bias by placing emphasis on high-frequency effects. They do have the advantage of not requiring any choice of smoothing parameter, and having much smaller bias than $\hat{\sigma}_\alpha^2$, vanishingly small in large samples. However it must be borne in mind that short-range serial correlation in the data will introduce serious bias into $\hat{\sigma}_R^2$ and (even more) into $\hat{\sigma}_{GSJ}^2$, while hardly affecting $\hat{\sigma}_\alpha^2$ at all.

3.5 Weighted smoothing

Up to now, the penalty term $\int g''^2$ has always been added to the ordinary residual sum of squares $\sum\{Y_i - g(t_i)\}^2$. In this section, we consider a more general form, in which the residuals are weighted. Suppose that $w_1, \ldots, w_n$ are strictly positive weights, and define the weighted residual sum of squares to be

$$\sum_{i=1}^{n} w_i\{Y_i - g(t_i)\}^2. \tag{3.20}$$

There is a number of contexts in which it is appropriate to assess the fit of the curve g to the points (t_i, Y_i) by a weighted residual sum of squares. The obvious application is to data where the Y_i are distributed with mean $g(t_i)$ but the variances of the Y_i are not equal. In this case it is natural to set the weights to be inversely proportional to the variances of the observations. However, we shall see in Chapter 5 below that smoothing based on the weighted residual sum of squares has far wider potential applicability.

3.5.1 *Basic properties of the weighted formulation*

In the subsequent discussion, define the matrix W to be the diagonal matrix with diagonal elements w_i. Given any function g in $\mathcal{S}_2[a, b]$, define the penalized weighted sum of squares $S_W(g)$ by

$$S_W(g) = \sum_{i=1}^{n} w_i\{Y_i - g(t_i)\}^2 + \alpha \int g''^2, \tag{3.21}$$

where, as usual, α is a positive smoothing parameter. Let $\hat{g}$ now be the minimizer of $S_W(g)$. The results of Section 2.3 can all be easily extended to apply to the weighted formulation. The result corresponding to Theorem 2.4 is the following.

Theorem 3.2 *Suppose $n \geq 3$ and that $a < t_1 <\ldots< t_n < b$. Suppose that the smoothing parameter α and the weights $w_i, i = 1, \ldots, n$ are all strictly positive. Given data values $Y_1, \ldots, Y_n$, the penalized weighted sum of squares $S_W(g)$ is uniquely minimized over g in $\mathcal{S}_2[a, b]$ by the natural cubic spline with knots at the points t_i having*

$$\mathbf{g} = (W + \alpha K)^{-1} W\mathbf{Y}. \tag{3.22}$$

Proof. The proof is exactly parallel to the argument set out in Sections 2.3.1 and 2.3.2, replacing the residual sum of squares by the weighted residual sum of squares. In Section 2.3.2, we replace $(\mathbf{Y} - \mathbf{g})^T(\mathbf{Y} - \mathbf{g})$ by $(\mathbf{Y} - \mathbf{g})^T W(\mathbf{Y} - \mathbf{g})$, and make consequential changes thereafter, leading to the expression (3.22). The details are left as an exercise for the reader.

Just as in the unweighted case, it is not appropriate to use (3.22) directly for calculation, and in the next section we set out the extension of the Reinsch algorithm to incorporate weights.

3.5.2 The Reinsch algorithm for weighted smoothing

It is easy to modify the derivation of the Reinsch algorithm in Section 2.3.3 to incorporate the weight matrix W. From (3.22) we have

$$\mathbf{g} = \mathbf{Y} - \alpha W^{-1} Q R^{-1} Q^T \mathbf{g},$$

and hence

$$\mathbf{g} = \mathbf{Y} - \alpha W^{-1} Q \gamma. \tag{3.23}$$

As before, substituting $Q^T\mathbf{g} = R\gamma$ we obtain, after some manipulation,

$$(R + \alpha Q^T W^{-1} Q)\gamma = Q^T \mathbf{Y}. \tag{3.24}$$

Because W is a strictly positive-definite diagonal matrix, the matrix $(R + \alpha Q^T W^{-1} Q)$ is a band matrix with $k = 2$ and has a Cholesky decomposition LDL^T where, as before, L is a lower diagonal band matrix with unit diagonal and D is a strictly positive diagonal matrix. The resulting algorithm, all of whose steps can be performed in $O(n)$ algebraic operations, can now be set out.

Algorithm for weighted spline smoothing

Step 1 Evaluate the vector $Q^T\mathbf{Y}$, by using the formula (2.7).

Step 2 Find the non-zero diagonals of $R + \alpha Q^T W^{-1} Q$, and its Cholesky decomposition factors L and D.

Step 3 Write (3.24) as $LDL^T\gamma = Q^T\mathbf{Y}$ and solve this equation for γ by forward and back substitution.

Step 4 From (3.23), use $\mathbf{g} = \mathbf{Y} - \alpha W^{-1} Q\gamma$ to find $\mathbf{g}$.

3.5.3 *Cross-validation for weighted smoothing*

The method of cross-validation is easily extended to deal with the weighted case. As usual, let $\hat{g}^{(-i)}(t_i)$ be the curve estimated omitting the data point (t_i, Y_i). The weights on all the other observations are assumed to be left alone. It is natural to calculate the cross-validation score taking the weights into account, so that more account is taken of points with high weight, giving the cross-validation function

$$CV(\alpha) = \sum_{i=1}^{n} w_i\{Y_i - \hat{g}^{(-i)}(t_i; \alpha)\}^2, \tag{3.25}$$

which would, as usual, be minimized to give the choice of α.

Just as in the unweighted case, it is not in fact necessary to solve n smoothing problems to find $CV(\alpha)$. Let $A_W(\alpha)$ be the hat matrix for the weighted case. Straightforward manipulations from (3.23) and (3.24) show that

$$I - A_W(\alpha) = \alpha W^{-1} Q(R + \alpha Q^T W^{-1} Q)^{-1} Q^T. \tag{3.26}$$

By exactly the same arguments as in the proof of Theorem 3.1 the cross-validation score can be written

$$CV(\alpha) = \sum_{i=1}^{n} w_i \left(\frac{Y_i - \hat{g}(t_i)}{\{I - A_W(\alpha)\}_{ii}} \right)^2, \tag{3.27}$$

where $\hat{g}$ is the minimizer of the penalized weighted sum of squares based on the full data set. By making minor modifications to Section 3.2.2 above to incorporate the diagonal matrix W where necessary, the diagonal elements of $I - A_W(\alpha)$ can be found from (3.26) in linear time, using the algorithm set out there to find the central diagonals of the matrix $(R + \alpha Q^T W^{-1} Q)^{-1}$.

3.5.4 Tied design points

One obvious application of the weighted formulation is to the case where there is more than one observation at each distinct point t_i, or equivalently there are ties among the original design points. Suppose that we have observation points $t_1 < t_2 < \ldots < t_n$ and that at the point t_i, independent observations $Y_{ij}, j = 1, \ldots, m_i$ are taken all with mean $g(t_i)$. Let $\bar{Y}_i$ be the mean of the observations at t_i,

$$\bar{Y}_i = m_i^{-1} \sum_{j=1}^{m_i} Y_{ij}.$$

Let $S(g)$ be the penalized sum of squares constructed from the original data

$$S(g) = \sum_i \sum_j \{Y_{ij} - g(t_i)\}^2 + \alpha \int g''^2; \tag{3.28}$$

then the problem of minimizing $S(g)$ is easily shown to be equivalent to that of minimizing the penalized weighted sum of squares

$$S_W(g) = \sum_i m_i \{\bar{Y}_i - g(t_i)\}^2 + \alpha \int g''^2. \tag{3.29}$$

It should be noted that although the problems of minimizing (3.28) and (3.29) are equivalent, there is a difference to be taken into account if cross-validation scores are calculated. The weighted formulation of cross-validation discussed in Section 3.5.3 above would correspond to leaving out all the observations at each t_i simultaneously. A more natural way of applying cross-validation to (3.28) is to leave out each of the observations Y_{ij} individually.

To construct such a score, let $\hat{g}^{(-ij)}$ be the minimizer of the penalized sum of squares constructed from all the data omitting the observation Y_{ij}. Let A_W be the hat matrix corresponding to the weighted formulation (3.29). Let $N = \sum m_i$, the total number of data points, and let S_i^2 be the sum of squares about their mean of the observations at t_i,

$$S_i^2 = \sum_{j=1}^{m_i} (Y_{ij} - \bar{Y}_i)^2.$$

The natural cross-validation score for the 'tied design point' case is

$$CV_T(\alpha) = N^{-1} \sum_{i=1}^{n} \sum_{j=1}^{m_i} \{Y_{ij} - \hat{g}^{(-ij)}(t_i)\}^2. \tag{3.30}$$

As usual, there is a computationally simpler version of (3.30). The natural modifications of the usual arguments show that the deleted residual can be expressed as

$$Y_{ij} - \hat{g}^{(-ij)}(t_i) = \frac{Y_{ij} - \hat{g}(t_i)}{1 - m_i^{-1}(A_W)_{ii}}. \tag{3.31}$$

Substituting (3.31) into (3.30) and simplifying yields

$$CV_T(\alpha) = N^{-1} \sum_{i=1}^{n} \frac{m_i\{\bar{Y}_i - \hat{g}(t_i)\}^2 + S_i^2}{\{1 - m_i^{-1}(A_W)_{ii}\}^2}. \tag{3.32}$$

We have already described in Section 3.5.3 the way in which the diagonal of A_W can be found in $O(n)$ operations. Hence, once the quantities $\bar{Y}_i$ and S_i^2 have been found, only $O(n)$ operations are required to find the function $CV_T(\alpha)$ for each α.

One interesting property of the methods described in this section is that they provide formulae that are continuous in the design points t_i. Suppose, for example, that by rounding off the t_i, additional ties are introduced. If the formula (3.32) is used for the cross-validation score, then individual values of $CV(\alpha)$ will only be altered by amounts due to the rounding. The coalescing of previously distinct points into tied design points will not, of itself, have any effect.

We return to problems involving weights and ties in the design points in Section 4.3. There we will adopt a more unified approach using a matrix notation.

3.6 The basis functions approach

In this section, we return to the fundamental problem of computing the minimizer of the penalized sum of squares $S(g)$. The discussion up to now, particularly in Section 2.3, has been based on the use of a particular roughness penalty, $\int g''^2$, that is amenable to a very complete and elegant mathematical analysis. Its properties reduce the problem of choosing the smoothing or interpolating g from being infinite-dimensional to finite-dimensional. The quadratic form $S(g)$ over a function space is replaced by a vector quadratic form such as that given in (2.10), and the smoothing problem can then be solved by linear algebra, for example through the Reinsch algorithm. In a certain sense, the nonparametric regression problem becomes a parametric one, though this is not the appropriate way of looking at it because there are essentially as many 'parameters'—the elements of the vector **g**—as observations. We shall see in Chapter 7 that these remarks extend to a wider range of roughness penalties, both in the univariate and the multivariate case, though it is not always possible to use band matrix methods to solve the linear equations in $O(n)$ operations.

However, this type of approach may be unsatisfactory for one of three reasons:

- the quantity of data may be so large that the systems of linear equations for the exact solution are too expensive to solve, and a cheaper approximation may be preferred; computational expense is not likely to be a problem with the Reinsch algorithm, but in very large systems it can become numerically unstable, particularly if the points t_i are irregularly distributed;

- there may be difficulties in interpreting a curve or surface g that requires a large number of parameters in its description; or

- there may be a preference for some other form of roughness penalty for which the required analysis is not available.

In any of these situations, one solution is to deliberately *impose* a finite-dimensional structure on the problem, by restricting the choice of g to the span $\mathcal{S}_B$ of a prescribed set of basis functions, $\beta_1, \ldots, \beta_q$, say. Thus we only consider functions g that can be expanded in the form

$$g(t) = \sum_{j=1}^{q} \delta_j \beta_j(t) \tag{3.33}$$

for some numbers $\delta_1, \ldots, \delta_q$. Of course, if the roughness penalty is $\int g''^2$ and the basis functions span the space of natural cubic splines with knots at the data points, then we would get the same solution as before. In other cases, we will be imposing a genuine constraint by restricting attention to g in $\mathcal{S}_B$, and the minimizer of $S(g)$ over $\mathcal{S}_B$ will not be identical to the minimizer over the space of all smooth functions. However it is intended that any difference will not be statistically important. The basis functions will be chosen so that their span includes good approximations to most smooth functions.

One popular choice for the basis functions is the set of cubic B-splines on a fixed grid of knots $s_1 < s_2 < \ldots < s_q$, usually taken to be equally spaced to cover the range of the points t_i. The B-splines form a set of natural cubic splines that are non-negative and have only limited support: for $3 \leq j \leq q-2$ the function β_j is zero outside (s_{j-2}, s_{j+2}), whilst β_1, β_2, β_{q-1} and β_q are similar, but linear outside (s_1, s_q).

Another possible approach, particularly appropriate if the function g is naturally required to be periodic, is to expand g in terms of a basis of trigonometric functions.

3.6.1 *Details of the calculations*

We can now set out the details of the calculations required for the basis functions approach. Consider the penalized weighted sum of squares

$$S_W(g) = \sum_{i=1}^{n} w_i\{Y_i - g(t_i)\}^2 + \alpha J(g), \tag{3.34}$$

where $J(g)$ is a quadratic roughness functional, such as $\int g''^2$. We seek a function g of the form (3.33) to minimize $S_W(g)$. In order to minimize the unweighted penalized sum of squares $S(g)$, proceed in exactly the same way setting the diagonal matrix W of weights to be the identity.

Since $J(g)$ is quadratic, there will be a $q \times q$ matrix K such that, for any g of the form (3.33),

$$J(g) = \boldsymbol{\delta}^T K \boldsymbol{\delta},$$

where $\boldsymbol{\delta}$ is the q-vector of coefficients δ_i. For example, if $J(g) = \int g''^2$, then

$$K_{jk} = \int \beta_j''(t)\beta_k''(t)dt.$$

If we let X denote the $n \times q$ matrix with $X_{ij} = \delta_j(t_i)$, then (3.34) can then be re-expressed as

$$S_W(g) = (\mathbf{Y} - X\boldsymbol{\delta})^T W(\mathbf{Y} - X\boldsymbol{\delta}) + \alpha\boldsymbol{\delta}^T K\boldsymbol{\delta}.$$

By standard arguments that are by now familiar, this quadratic form is minimized by

$$\hat{\boldsymbol{\delta}} = (X^T WX + \alpha K)^{-1} X^T W\mathbf{Y}. \tag{3.35}$$

Of course, details of the efficient numerical evaluation of $\hat{\boldsymbol{\delta}}$ will depend on the structure of K and X. Notice that there is no need to take explicit note of ties among the points t_i, nor is the order of the t_i relevant. Furthermore, once the matrix $X^T WX$ and the vector $X^T W\mathbf{Y}$ have been calculated, the linear system to be solved is of size q rather than n.

If the B-spline basis is used and $J(g) = \int g''^2$ then the matrices X and K are both banded, so that $X^T WX$ and $X^T W\mathbf{Y}$ can be found in $O(n)$ operations and the equations (3.35) then solved in $O(q)$ operations. As well as being economical, this algorithm is also stable numerically. In fact, if the knots s_j are taken to be the ordered distinct values among $\{t_i\}$, it is a rival to the Reinsch algorithm for computing the ordinary cubic smoothing spline precisely; this is the approach used by the S routines described in Section 8.1 below. Although the workload is still linear in n (once the $\{t_i\}$ are ordered) the operation count for the B-spline method is rather larger than for the Reinsch algorithm, but it does achieve greater numerical stability and therefore accuracy. In our experience, however, the difference in accuracy is not material in the context of statistical regression analysis, with sensible values of α and moderately evenly spaced t_i.

3.7 The equivalent kernel

3.7.1 Roughness penalty and kernel methods

The roughness penalty approach set out in this book is of course only one of a number of curve estimation procedures that are available. The main conceptual advantage of the roughness penalty method is that it allows explicit specification of the way in which goodness-of-fit to the data is to be measured. So far, we have concentrated attention on penalizing the residual sum of squares $\sum\{Y_i - g(t_i)\}^2$ for roughness, but we shall see below that the approach is applicable to a much wider range of possibilities. Of course, a price that has to be paid for this versatility is that the estimator is obtained by solving a minimization problem, rather than by calculating an explicit formula.

Weight function, or kernel, methods start from a different point of view, and in general define the estimate at each point t as being an explicit function, usually a weighted average, of 'local' observations Y_i. We shall not discuss kernel methods in detail, but refer readers to other texts that concentrate on them, such as Härdle (1990). An interesting exposition of some subtle issues to do with the way in which kernel estimates can be constructed is given by Chu and Marron (1992); see the contribution to the discussion by Silverman (1992) for some remarks about roughness penalty methods in this context.

3.7.2 Approximating the weight function

It is interesting to note that there is a relationship between the spline smoothing estimate and a particular kernel estimate that may be of some conceptual value in providing a deeper understanding of the spline smoothing method. For fuller details of the discussion of this section, see Silverman (1984a).

It follows immediately from the quadratic nature of the penalized sum of squares $S(g)$ that the spline smoother $\hat{g}$ is linear in the observations Y_i, in that there exists a *weight function* $G(s,t)$ such that the estimate can be written

$$\hat{g}(s) = n^{-1} \sum_{i=1}^{n} Y_i G(s, t_i). \tag{3.36}$$

For each t_i, the weight function $G(\cdot, t_i)$ is the curve estimate that would

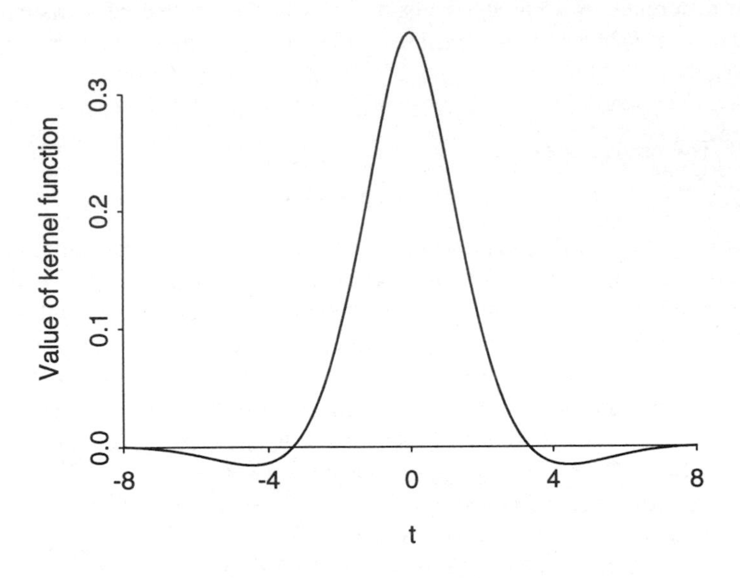

Figure 3.1. *The effective kernel function* κ

be obtained—with the same value of α—if $Y_i = n$ and $Y_j = 0$ for $j \neq i$. Thus the weight function may be thought of as a smoothed-out version of a delta-function at t_i. The weight function clearly depends on all the design points $t_1, \ldots, t_n$ and on the smoothing parameter, but it does not depend on the observation values Y_i.

It is possible to show that the weight function has an approximate form that is of some intuitive interest. Suppose that n is large, and that the design points have local density $f(t)$. Provided s is not too near the edge of the interval on which the data lie, and α is not too big or too small, we then have the approximation

$$G(s,t) \approx \frac{1}{f(t)} \frac{1}{h(t)} \kappa\left(\frac{s-t}{h(t)}\right). \tag{3.37}$$

The *kernel function* κ is plotted in Figure 3.1 and is given by the formula

$$\kappa(u) = \frac{1}{2} \exp\left(-\frac{|u|}{\sqrt{2}}\right) \sin\left(\frac{|u|}{\sqrt{2}} + \frac{\pi}{4}\right). \tag{3.38}$$

The *local bandwidth* $h(t)$ satisfies

$$h(t) = \alpha^{\frac{1}{4}} n^{-\frac{1}{4}} f(t)^{-\frac{1}{4}}. \tag{3.39}$$

If we examine the approximation (3.37) in detail we can see that the spline smoother is approximately a convolution—or weighted moving average—smoothing method. Nearby observations Y_i contribute most to the estimate, and the speed with which the influence of data points dies away is governed by the value (3.39) of the local bandwidth $h(t)$. The following general comments can be made:

- The form of κ demonstrates that the observation at t_i has an influence on nearby parts of the curve $\hat{g}$ that dies away exponentially—a favourable comparison with some other curve fitting methods such as polynomial regression. A refinement of the approximation near the boundary of the interval on which the data lie (see Silverman, 1984a) shows that the weight function is distorted there, though its exponential decay away from the boundary is not affected.

- Altering the smoothing parameter α affects the local bandwidth in the same multiplicative way everywhere. Note that the local bandwidth is proportional to the fourth root of α. We should not be surprised if the appropriate numerical value of α varies widely between different problems, particularly if the scale of the design variable is different.

- The dependence of the local bandwidth on the local density f of data points is intermediate between fixed-width convolution smoothing (no dependence on f) and smoothing based on an average of a fixed number of neighbouring points (effective local bandwidth proportional to $1/f$). Theoretical considerations discussed in Silverman (1984a) suggest that such intermediate behaviour is desirable, and that it will provide almost ideal adaptivity to effects caused by variability in the density of design points t_i.

In conclusion, it should be stressed that the importance of the equivalent kernel formulation is conceptual, in helping to give intuition about what the spline smoother actually does to the data. It should go almost without saying that it is not intended to be used for calculation!

3.8 The philosophical basis of roughness penalties

In this section, we expand on the discussion of Section 1.2.1 and set out a number of approaches that have been used to motivate the roughness penalty approach to nonparametric smoothing, and discuss some ramifications of them. The various approaches have a very long history indeed, dating back beyond Whittaker (1923). For some historical remarks, see Whittle (1985).

3.8.1 Penalized likelihood

The idea set out in Section 1.2.1 was the direct concept that there are two aims in curve estimation, which to some extent conflict with one another, to maximize goodness-of-fit and to minimize roughness. Penalizing the residual sum of squares by adding a roughness penalty term is an obvious way of making the necessary compromise explicit. The approach can be seen as a particular case of the more general concept of *penalized likelihood*, first discussed in the modern literature by Good and Gaskins (1971).

To describe the penalized likelihood approach in general, suppose that the distribution of a set of observed data $\mathbf{X}$ is governed by a curve g, say. In Good and Gaskins (1971) attention is concentrated on the case where the data are independent observations with underlying probability density function g. Let $\ell(g)$ be the log-likelihood of g given the data—in the probability density estimation context $\sum \log g(X_i)$. In the regression case, if it is assumed that the data are independently normally distributed with means $g(t_i)$ and equal variances σ^2, then the log-likelihood is given (up to a constant) by

$$\ell(g) = -\frac{1}{2\sigma^2}\sum_i \{X_i - g(t_i)\}^2.$$

The unconstrained maximization of $\ell(g)$ does not provide a sensible estimate of g. In the case of probability density estimation as discussed by Good and Gaskins (1971, 1980), the likelihood is unbounded above, and tends to infinity as g approaches a sum of delta functions at the data points. In the regression case, the likelihood is maximized by any curve that interpolates the data. Arguing from a Bayesian point of view that will be discussed in more detail in Section 3.8.3, Good and Gaskins (1971) suggested subtracting from the log-likelihood a roughness penalty or, in their terminology, a 'flamboyancy functional', that measures the local variation in g. In the regression context, if the roughness penalty is $\frac{1}{2}\lambda \int g''^2$, then the penalized likelihood is equal to

$$\ell_P(g) = -\frac{1}{2\sigma^2}\sum_i \{X_i - g(t_i)\}^2 - \tfrac{1}{2}\lambda \int g''^2. \tag{3.40}$$

If the parameter λ is set to α/σ^2, where α is a smoothing parameter, then it is immediate that the maximization of ℓ_P is precisely equivalent to the minimization of the penalized sum of squares $S(g)$ as defined in (1.3).

3.8.2 *The bounded roughness approach*

The motivation of Reinsch (1967), in discussing spline smoothing, was somewhat different. Suppose we are trying to fit a curve g to a set of observations Y_i in the usual regression context. A possible way of avoiding the difficulties that arise if the residual sum of squares $\sum\{Y_i - g(t_i)\}^2$ is maximized in an unconstrained way is to consider the *constrained* optimization problem

$$\min_g \sum\{Y_i - g(t_i)\}^2 \text{ subject to } \int g''^2 \le C. \tag{3.41}$$

A standard Lagrangian argument shows that the function g that solves the minimization problem (3.41) can be found by adding a Lagrange term $\alpha \int g''^2$ to the residual sum of squares and then minimizing in an unconstrained way. This will give exactly the penalized sum of squares $S(g)$, with smoothing parameter equal to the Lagrange multiplier α. In order to solve (3.41) for a particular C, it is necessary to search on α until the optimizing function $\hat{g}$ satisfies the constraint $\int g''^2 = C$. Since $\int \hat{g}''^2$ can easily be shown to be a decreasing function of α, this search is not prohibitively expensive since it can be carried out by a binary search procedure. However, it is unusual for the value C to be directly meaningful, and the usual practical approach in statistics is to regard the Lagrange multiplier α as the controlling parameter for the smoothing method.

3.8.3 *The Bayesian approach*

It is relatively rare for statistical writers to justify penalized least squares or maximum likelihood via a constrained maximum likelihood argument. Those who are not satisfied with regarding the method as an attractive *ad hoc* device usually appeal to a Bayesian justification. For example, Whittle (1985) wrote

> It is plain that one cannot discuss these matters without being prepared to consider a Bayesian formulation (which I am understanding in a frequentist non-personal sense) ... it should be recognized that, with the exception of a short historical interlude, the [Bayesian] approach has always been considered a perfectly natural one.

Intuitively speaking, the Bayesian justification of penalized maximum likelihood is to place a prior density proportional to $\exp(-\frac{1}{2}\lambda \int g''^2)$ over the space of all smooth functions. In a certain sense that we shall discuss below, the larger the value of λ, the more weight is put on functions with smaller roughness. With this prior, the posterior log density of the function g is then, in the regression context, equal to $\ell_P(g)$ as defined in

(3.40) above, and so the spline smoother $\hat{g}$ is the posterior mode given the data. Wahba (1978, 1983), drawing on earlier work of Kimeldorf and Wahba (1970), developed this approach in regression, and suggested the use of pointwise error bands for the curve estimate based on the posterior distribution. We now summarize some results of these papers in the next section.

A Gaussian process prior

Both the prior and the posterior log densities are quadratic forms in the function g, and so they correspond to a Gaussian process structure. The prior distribution is 'partially improper', in that it is invariant under the addition of a constant or linear function to g. One way of visualizing the prior distribution is to write

$$g(t) = A + Bt + \lambda^{-\frac{1}{2}} \int_0^t W(s)ds \tag{3.42}$$

where A and B have improper uniform distributions on $(-\infty, \infty)$ and $W(s)$ is a Brownian motion on $(-\infty, \infty)$.

Since the posterior distribution is also a Gaussian process, the estimate $\hat{g}$ is the posterior mean as well as being the posterior mode. It can be shown that the posterior distribution of the vector $\mathbf{g}$ of values $g(t_i)$ is multivariate normal with variance matrix equal to $\sigma^2 A(\alpha)$, where $\alpha = \sigma^2\lambda$ as above, and $A(\alpha)$ is the hat matrix as defined in Section 3.2.1. Thus a 95% Bayesian posterior probability interval for each $g(t_i)$ is given by $\hat{g}(t_i) \pm 1.96\sigma A(\alpha)_{ii}$, giving another application of the algorithm described in Section 3.2.2 for finding the diagonal elements of the hat matrix.

In practice, the standard deviation σ is usually estimated from the data, for example using one of the methods discussed in Section 3.4 above. The recommendation in Wahba's papers is to choose the smoothing parameter α by GCV.

An example of inference regions obtained by a minor modification of this approach is given in Figure 3.2, taken from Silverman (1985). The measurements plotted in this figure are the logarithms (to base 10) of the population count per millilitre of the organism *Staphylococcus aureus* in a heart infusion broth. They were taken in a microbiological experiment carried out at the ARC Meat Research Institute, Langford, Bristol. The pointwise 95% probability intervals are calculated at each t_i and interpolated linearly between these points.

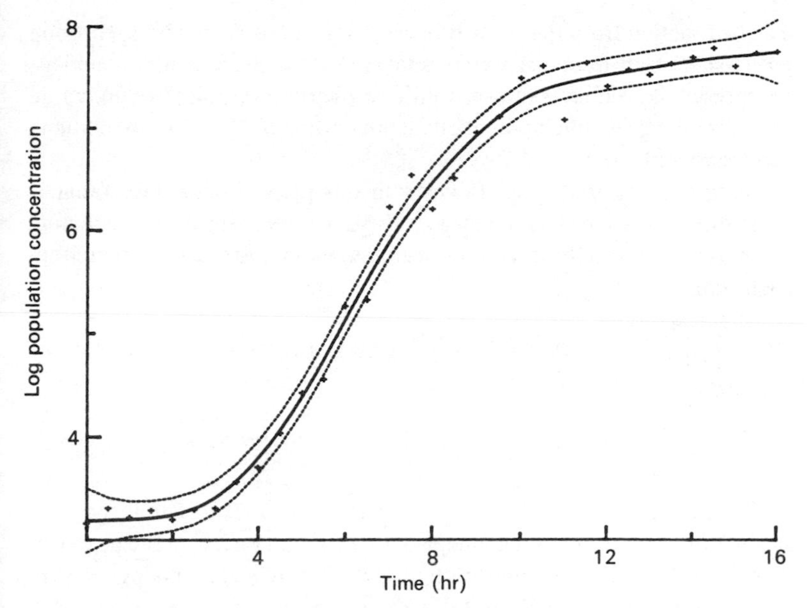

Figure 3.2. *Microbiological data, with estimated growth curve and probability intervals. Reproduced from Silverman (1985) with the permission of the Royal Statistical Society.*

A paradox

The space of curves g within which we wish to choose a regression fit to the data is, of course, infinite dimensional, even if we restrict attention to curves on a bounded interval $[a, b]$ that are smooth in some sense, for example having $\int_a^b g''^2$ finite. This formulation leads to a paradox, alluded to by Wahba (1983).

To set out this paradox, suppose that the interval $[a, b]$ is $[0, 1]$, and let $\mathcal{S}$ be the space of functions on $[0, 1]$ for which $\int_0^1 g''^2$ is finite. There is an infinite sequence ϕ_ν of orthonormal functions and an increasing sequence of eigenvalues ρ_ν, with $0 = \rho_1 = \rho_2 < \rho_\nu$ for $\nu \geq 3$, such that

- every function g in $\mathcal{S}$ can be expanded in the form $\sum g_\nu \phi_\nu$ for suitable coefficients g_ν; and
- $\int_0^1 g''^2 = \sum \rho_\nu g_\nu^2$ for g in $\mathcal{S}$.

The prior density (up to a constant of proportionality) $\exp(-\frac{1}{2}\lambda \int g''^2)$ can be written as

$$\exp(-\tfrac{1}{2}\lambda \sum_{\nu \geq 3} \rho_\nu g_\nu^2) = \prod_{\nu \geq 3} \exp(-\tfrac{1}{2}\lambda \rho_\nu g_\nu^2).$$

Thus a function from the prior process can be constructed by specifying its expansion in terms of the functions ϕ_ν: the coefficients g_1 and g_2 would each have improper prior uniform distributions on $(-\infty, \infty)$, while for $\nu \geq 3$ the g_ν would have independent normal distributions with mean zero and variances $(\lambda\rho_\nu)^{-1}$.

Suppose, now, that g is drawn from this prior distribution. What is the distribution of the roughness $\int g''^2$? We have, writing $N(0, \cdot)$ to denote normal random variables with the specified variances independent within each sum,

$$\begin{aligned}\int g''^2 &= \sum_{\nu=3}^{\infty} \rho_\nu g_\nu^2 = \sum_{\nu=3}^{\infty} \rho_\nu N(0, (\lambda\rho_\nu)^{-1})^2 \\ &= \lambda^{-1} \sum_{\nu=3}^{\infty} N(0,1)^2 = +\infty \text{ with probability 1.}\end{aligned}$$

Thus, although we set out to construct a prior distribution over the space $\mathcal{S}$ of smooth functions, the prior distribution (and also the posterior distribution) is entirely supported *outside* this space! This does not, in itself, contradict the result that $\hat{g}$ is the posterior mean, because the operation of taking the expectation over the posterior is, in a certain sense, a smoothing operation that yields a smooth mean $\hat{g}$ even though any individual realization from the posterior would not have finite smoothness. However, it demonstrates that a careful limiting argument would be needed to justify the Bayesian inference in the infinite dimensional space $\mathcal{S}$.

One possible approach might be to perform the inference on the space spanned by the first N functions ϕ_ν, and let N tend to infinity. For any finite N the argument given above shows that the prior distribution of roughness is χ^2_{N-2}/λ, according with the notion that more weight is put on functions with lower smoothness if λ is large. However it would still be the case that, in the limit, the prior and posterior probabilities that $\int g''^2$ is finite is zero and so the normalizing constant in the likelihood would be zero.

3.8.4 *A finite-dimensional Bayesian formulation*

Another way of resolving the paradoxes and difficulties involved in the infinite-dimensional Bayesian formulation is to work on a fixed finite-dimensional space rich enough to allow the inferences set out by Wahba to be exactly correct. See Silverman (1985) for any details not given here.

Given distinct points $t_1, \ldots, t_n$, let $\mathcal{S}_{NCS}$ be the space of natural cubic splines with knots at the t_i. An obvious way of parametrizing the space

$\mathcal{S}_{NCS}$ is by the vector $\mathbf{g}$ of values $g(t_i)$, which will specify a natural cubic spline g uniquely. Using $\stackrel{c}{=}$ to denote equality up to a constant, take the prior log density over $\mathcal{S}_{NCS}$ to be

$$\ell_{prior}(g) \stackrel{c}{=} -\tfrac{1}{2}\lambda \int g''^2 = -\tfrac{1}{2}\lambda \mathbf{g}^T K \mathbf{g}, \tag{3.43}$$

defining the matrix K as in Theorem 2.1. The matrix K is singular, with rank $n-2$. Its two zero eigenvalues correspond to constant and linear functions g. Were it not for these the prior would be multivariate normal over $\mathbf{g}$ with covariance matrix K^{-1}; as it is, the prior is 'partially improper' with infinite variance given to these two eigenvectors of K.

Assuming that the observations Y_i are independently normally distributed with means $g(t_i)$ and variance σ^2, the posterior log-likelihood will then be (allowing the constant in $\stackrel{c}{=}$ to depend on the data)

$$\begin{aligned} \ell_{post}(g) &\stackrel{c}{=} -\tfrac{1}{2}\lambda \mathbf{g}^T K \mathbf{g} - \tfrac{1}{2}\sigma^{-2}(\mathbf{Y}-\mathbf{g})^T(\mathbf{Y}-\mathbf{g}) \\ &\stackrel{c}{=} -\tfrac{1}{2}\sigma^{-2}\{\mathbf{g}^T A(\alpha)^{-1}\mathbf{g} - 2\mathbf{g}^T\mathbf{Y}\}. \end{aligned} \tag{3.44}$$

It follows from (3.44) that the posterior distribution of $\mathbf{g}$ will be multivariate normal with mean $\hat{\mathbf{g}} = A(\alpha)\mathbf{Y}$, and variance matrix $\sigma^2 A(\alpha)$, just as in the infinite-dimensional case discussed above. The posterior mean curve will be precisely the element of $\mathcal{S}_{NCS}$ satisfying $\mathbf{g} = A(\alpha)\mathbf{Y}$, namely the spline smoother $\hat{g}$.

3.8.5 Bayesian inference for functionals of the curve

Very many important questions in curve estimation involve quantities such as the gradient or maximum of g rather than the curve g itself. The Bayesian formulation allows posterior probability distributions to be found for any functional of g that is well defined for g in $\mathcal{S}_{NCS}$. The basic approach depends largely on whether the functional of interest is linear or nonlinear.

Linear functionals

Consider, first of all, the estimation of a linear functional $\psi(g)$ such as $g'(t)$ for a particular t. Define constants ψ_i such that

$$\psi(g) = \sum \psi_i g_i \text{ for any } g \text{ in } \mathcal{S}_{NCS}.$$

Then, by standard properties of the multivariate normal distribution, the posterior distribution of $\psi(g)$ will be normal with mean $\boldsymbol{\psi}^T A(\alpha)\mathbf{Y} = \psi(\hat{g})$, and variance $\sigma^2 \boldsymbol{\psi}^T A(\alpha)\boldsymbol{\psi}$.

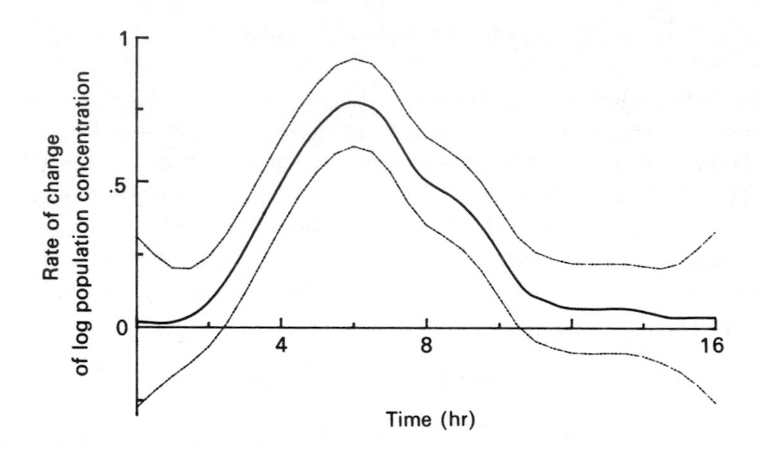

Figure 3.3. *Estimated growth rate for microbiological data, with probability intervals. Reproduced from Silverman (1985) with the permission of the Royal Statistical Society.*

For each i, the quantity ψ_i can be found by evaluating the functional ψ on the natural cubic spline satisfying $g(t_i) = 1$ and $g(t_j) = 0$ for $j \neq i$. Alternatively it may be helpful to use a different basis for S_{NCS}, such as B-splines. For details of the calculations in this case, see Silverman (1985).

In the growth curve example presented in Figure 3.2, the rate of growth was of particular interest. In Figure 3.3 the methodology of this section is used to produce an estimate of the growth rate g' together with pointwise 95% posterior probability intervals. It can be seen—as one would expect—that the growth rate cannot be considered to have been estimated with particular accuracy.

Nonlinear functionals: simulation from the posterior

Now suppose that $\psi(g)$ is a nonlinear functional, such as $\max g'(t)$, a quantity of interest in the growth curve example. In contrast to the linear case, the exact posterior distribution of $\psi(g)$ is unlikely to be tractable, but progress can be made by a Monte Carlo approach. Simulate curves g from the posterior distribution, by simulating vectors $\mathbf{g}$ from the multivariate normal distribution with mean $\hat{\mathbf{g}}$ and variance $\sigma^2 A(\alpha)$; for each realization $\mathbf{g}$ find the corresponding curve g and evaluate $\psi(g)$. By repeating a large number of times, a series of independent realizations from the posterior distribution of $\psi(g)$ may be found, and the posterior mean and any other statistics of the posterior may be estimated from these. Just as in the linear case, it may be more convenient to work in terms of a different parametrization of $\mathcal{S}_{NCS}$; see Silverman (1985).

As an illustration, this technique was applied to the estimation of the maximum growth rate for the microbiological data. A naive approach would estimate this quantity by the maximum value (0.77) of the solid curve in Figure 3.3. Simulation from the posterior indicates that the posterior mean is 0.84 and the standard deviation 0.06. The discrepancy between 0.84 and 0.77 is important, and not at all surprising; by Jensen's inequality one has

$$E_{post} \max_t g'(t) > \max_t E_{post} g'(t) = \max_t \hat{g}'(t).$$

In more informal terms, the solid curve in Figure 3.3 is the average of a population of curves, each of which will tend to have maximum value higher than the maximum of the mean curve.

Another example of the estimation of a nonlinear functional is described in the next section.

3.9 Nonparametric Bayesian calibration

An example of the estimation of a nonlinear functional arises in calibration, for a detailed recent survey of which see, for example, Osborne (1991). Suppose that g is a monotonically increasing function, that is estimated on the basis of a *calibration set* of observations Y_i taken at points t_i. The problem of calibration is the converse of the usual prediction problem. In its simplest form, the problem is then that of estimating the point τ at which the curve crosses a given level η, say, so that the nonlinear functional that we are trying to estimate is $g^{-1}(\eta)$. More realistically, one or more observations Y' are taken on a new individual, and on the basis of these observations it is of interest to estimate t for that individual. A natural, if naive, point estimate of t is $\hat{g}^{-1}(\bar{Y}')$, where $\hat{g}$ is some suitable estimate of g obtained from the data (t_i, Y_i), and $\bar{Y}'$ is the average of the observations taken on the new individual. But in assessing any kind of variability of this estimate it is necessary to take into account two sources of variability, that involved in the estimation of g, and that involved in the observation(s) Y'.

3.9.1 The monotonicity constraint

One interesting feature of the calibration problem is that it is natural to assume that the curve g is monotonically increasing. It is extremely easy to build this monotonicity requirement into the Bayesian framework. Let S_{MON} be the space of natural cubic splines with knots at the t_i that are monotonically increasing. It is easy to check whether a particular natural cubic spline g is monotonically increasing. A crude check, sufficient for most practical purposes, is to consider whether the sequence of values at the knots is increasing; more sensitively, the pieces of quadratics that make up the curve g' can be checked to ensure that g' is strictly positive on every interval (t_i, t_{i+1}).

We assume exactly the same prior log density (3.43) as before, but restrict to the space S_{MON}, setting the likelihood to be zero for non-increasing curves g. For all g in S_{MON}, the posterior likelihood is (3.44) as before, and so the posterior is now a *truncated* multivariate normal distribution. If—as almost invariably happens in practical calibration contexts—the spline smoother $\hat{g}$ based on the calibration data is increasing, then $\hat{g}$ will continue to be the mode of the posterior distribution, though it will of course no longer be the mean. Simulation from the posterior distribution over S_{MON} is, in principle, easily carried out by a rejection sampling approach; curves are simulated from the non-truncated multivariate normal posterior, but are only accepted if they fall in S_{MON}. Provided $\hat{g}$ is itself in S_{MON}, and the calibration data set does not display excessive variability, a reasonable proportion of the untruncated posterior normal distribution will fall in S_{MON} anyway, and the rejection sampling algorithm will be efficient enough for practical use.

In any case, non-monotonicity of $\hat{g}$ casts some doubt on the appropriateness of the experiment for calibration purposes. Sometimes the difficulty may be caused by a single value, perhaps an outlier, or an observation at a point with a high leverage value (particularly at or near the end of the range of the data). Such observations may not actually cause $\hat{g}$ to be non-monotonic but may cause the rejection sampling algorithm to have low acceptance probability. In these circumstances a more sophisticated simulation approach could be used, but it is probably easiest to use the *ad hoc* approach of downweighting any offending observations somewhat.

3.9.2 Accounting for the error in the prediction observation

If the calibration problem were merely that of inference for $g^{-1}(\eta)$ for some known η then we could proceed by simulating from the posterior over S_{MON}, and evaluating $g^{-1}(\eta)$ for each posterior realization, to build up a sample from the posterior distribution of $g^{-1}(\eta)$. But in practice the observation Y' is itself subject to error. A useful model is $Y' = \eta + \epsilon'$ where ϵ' has a $N(0, \sigma^2)$ distribution, and $\eta = g(\tau)$ for some value τ that it is of interest to estimate. The value of σ^2 will be the same as the error variance of the calibration experiment, and so can be estimated from the calibration data.

A simple Bayesian approach now proceeds as follows. Suppose η has a uniform prior distribution, so that the posterior distribution of η is $N(Y', \sigma^2)$. Realizations from the posterior distribution of τ, given the data in both parts of the experiment, can be obtained by repeatedly simulating g_{sim} from the posterior distribution of g and η_{sim} from the posterior distribution of η, and evaluating $g_{sim}^{-1}(\eta_{sim})$. It may be appropriate to restrict the prior for η to be uniform over a specified range, for example the image under $\hat{g}$ of a plausible range for t. The resulting truncation of the normal posterior distribution makes η_{sim} scarcely more difficult to obtain.

The validity of this approach depends on the prior for Y' being independent of that for g, rather than the more realistic approach of placing a prior over τ and mapping this to a prior for η. (Even the truncation approach described is not, strictly speaking, valid.) However, if (as is common in calibration experiments) the derivative g' varies only slowly, a uniform prior on τ will map to a prior on η that is locally nearly uniform, and hence the posterior for η will be close to that corresponding to a uniform prior on η.

3.9.3 Considerations of efficiency

In order to produce independent realizations of τ it is necessary to use independent realizations of η_{sim} and of g_{sim}. Because it may be much easier to generate η_{sim} than g_{sim}, it may be worth sacrificing independence in the posterior sample by using a number of independent η_{sim} for each g_{sim}. Osborne (1990) considers the implications of simulating M values η_{sim} for each of N curves g_{sim}, giving MN realizations from the posterior for τ. As long as N is large, these realizations should give good estimates of the quantities generally required, such as the mean, variance, and various quantiles of the posterior distribution.

The aim is to obtain the most accurate estimates of these quantities in a given amount of computer time. It turns out (Osborne, 1990, Theorem 3.1) that the best choice of M to do this depends not on the time available, but only on properties of the model, the time taken to generate single realizations, and on the particular posterior quantity of interest. Given

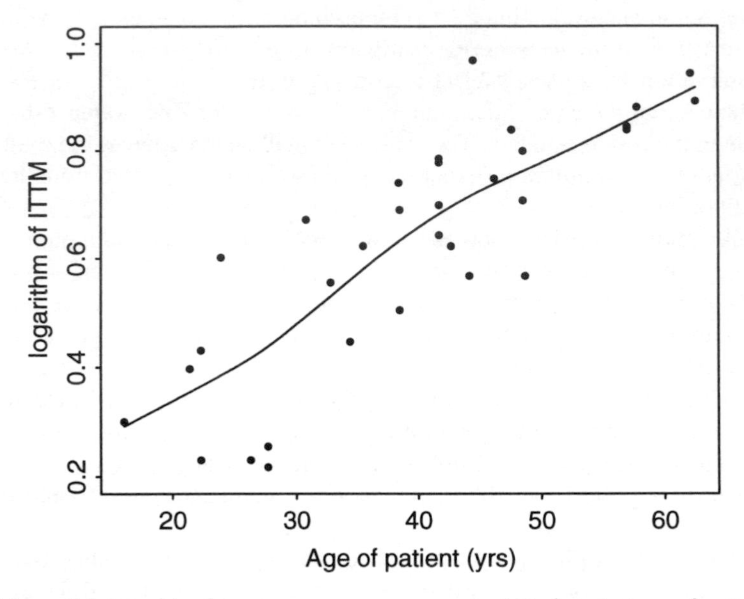

Figure 3.4. *Intact tooth transparency measurements plotted against age of patient for each of 32 upper central teeth. Nonparametric regression curve of* log *ITTM shown, α chosen by GCV.*

twice as much time, one should keep M the same and double N. For typical problems of interest, the optimal value of M is in the range 10 to 25. To get decent estimates of quantities such as 10% and 90% quantiles, it will be advisable to ensure that MN is then of the order of 10,000.

3.9.4 An application in forensic odontology

A specific example of a calibration problem is described in detail by Osborne (1990). Forensic odontologists are often called upon to infer the age of a victim (or sometimes a criminal) on the basis of teeth. The roots of all teeth contain dentine, some of which is transparent. The amount of transparent root dentine is known to increase with age, and is most conveniently quantified by the length measurement of intact tooth transparency (ITTM) in millimetres.

The Glasgow Dental Hospital supplied a number of teeth extracted from patients of known age t_i. For each of these teeth the ITTM was measured. Various types of tooth were considered (upper and lower; central, canine and lateral) and graphs of ITTM against age for all six possible types indicate that a logarithmic transformation is appropriate to stabilize the variance. We therefore let Y_i be the logarithm of the ITTM for tooth i. The most frequent type of tooth in the study was upper central, the data for which are plotted in Figure 3.4, together with the nonparametric regression curve with smoothing parameter chosen by GCV. It can be seen that the dependence of log ITTM on age is somewhat sigmoidal, as is typical of calibration experiments.

The fitted curves for the other types of tooth showed a variety of behaviour from almost linear to curves having higher slopes in different parts of the range in each case. All were clearly monotonic except for upper lateral teeth, where the curve was virtually constant over most of the range, indicating the unsuitability of this type of tooth for calibration purposes.

In order to validate the procedure, a further test set of teeth of various types from new patients (of known ages) was used. The measurements for the upper lateral teeth confirmed the futility of attempting to infer age from this type of tooth. For the other five types (44 teeth in all) the age of each patient was inferred by finding the median of the posterior distribution of τ, using the appropriate calibration sets and choosing the smoothing parameter by GCV. A slight modification to the fitting procedure was applied, as described below. The predictions were at least as good, and in the case of one type of tooth noticeably better, than those obtained by a standard parametric method. It is encouraging that the method works well even on such relatively small calibration data sets with appreciable scatter.

Correlation between teeth from the same patient

This data set has an interesting feature, of more general interest than its relevance to calibration. In some cases there are two teeth of a particular type from a single patient, and so the covariance structure of the calibration set needs some adjustment to account for this. Suppose that the distinct patients have ages $\tau_1, \ldots, \tau_m$, and write Y_{jk} for the log ITTM of the kth tooth contributed by the jth patient. Let $\bar{Y}_j$ be the average of the one or two observations for patient j. A reasonable model for the data is to assume that

$$Y_{jk} = g(\tau_j) + \epsilon_{jk}$$

where the ϵ_{jk} are all $N(0, \sigma^2)$ random variables, with correlations 0 between variables with differing values of j, and ρ between variables with the same value of j but different values of k (different teeth taken from the same patient). It is then easy to show that, if patient j contributes two teeth, then var $\bar{Y}_j = \frac{1}{2}\sigma^2(1 + \rho)$. For any given value of ρ, therefore, the spline smoother was constructed by minimizing

$$S_W(g) = \sum_j w_j\{\bar{Y}_j - g(\tau_j)\}^2 + \alpha \int g''^2 \tag{3.45}$$

where $w_j = 1$ if patient j contributed one tooth to the study and $2/(1+\rho)$ otherwise. This is a natural extension of the technique of Section 3.5.4 for data with tied design points.

Both σ^2 and ρ were estimated from the data, by a simple iterative approach. Initially ρ was set to 0, so that $w_j = 1$ or 2 for each j. The curve was then estimated by minimizing the weighted penalized residual sum of squares (3.45), choosing α by GCV, and an estimate $\hat{\sigma}^2$ was given by the weighted version of (3.19). For any patient contributing two teeth, $Y_{j1} - Y_{j2}$ has variance $2(1-\rho)\sigma^2$, so an estimate of ρ was obtained by setting

$$\hat{\rho} = 1 - \hat{\sigma}^{-2}(2n')^{-2}\sum{}'(Y_{j1} - Y_{j2})^2$$

where n' is the number of patients contributing two teeth to the study, and $\sum'$ denotes a sum over such patients. The estimate $\hat{\rho}$ of ρ was then used to update the weights w_j. Iterating this process to convergence gave estimates of σ^2 and ρ. For the upper central teeth shown in Figure 3.4, for example, the estimate of ρ was 0.50, and similar values were obtained for most of the other types of tooth. In each case, the posterior likelihood was taken to be $-S_W(g)/(2\sigma^2)$, with S_W defined as in (3.45) and the various parameters estimated from the data by the iterative procedure.

CHAPTER 4

Partial splines

4.1 Introduction

We have so far been concerned entirely with the dependence of our observations Y on a single explanatory variable t. Whilst this is sufficient to deal with a number of problems of interest, there are many situations in practice where observed responses are influenced simultaneously by *several* variables. Statistical analysis of the dependence on explanatory variables then usually leads to the use of *multiple regression*. The explanatory variables may be either quantitative (numerical) or qualitative (categorical) and the most well known general framework is provided by the *linear model*

$$Y_i = \mathbf{x}_i^T \beta + \text{ error.} \tag{4.1}$$

Here $\mathbf{x}_i$ is a vector of explanatory variables for the ith observation, and β the corresponding vector of regression coefficients, to be estimated. In general, the vector $\mathbf{x}_i$ may include a constant entry 1 for an intercept term, indicator variables to model categorical explanatory variables, and products of other components so that interactions can be assessed. The simple case (1.1) of univariate linear regression is given by setting $\mathbf{x}_i = (1, t_i)^T$ and $\beta = (a, b)^T$. It is not appropriate here to give a complete discussion of the linear model: the interested reader should consult one of the many standard textbooks on the subject.

What we shall be concerned with is relaxing the assumptions of linearity in equation (4.1), in much the same way that straight line regression (1.1) was generalized to curve fitting as in (1.2), in Chapter 2. The natural analogy would be to consider the model

$$Y_i = g(\mathbf{t}_i) + \text{ error,} \tag{4.2}$$

where g is an arbitrary, but smooth, real-valued function of a vector variable, but we shall not make such a complete generalization immediately. As we shall see in Chapter 7, nonparametric regression on several variables poses new conceptual problems, and a considerable computational burden. However, the machinery of penalized least squares and cubic splines presented in Chapters 2 and 3 is sufficient to provide a very useful generalization of (4.1), allowing just one explanatory variable to be treated in a nonparametric fashion.

4.2 The semiparametric formulation

Suppose that for each observation Y_i there are $p+1$ explanatory variables: a p-vector $\mathbf{x}_i$ and a scalar t_i. This chapter is concerned with models of the form

$$Y_i = \mathbf{x}_i^T \boldsymbol{\beta} + g(t_i) + \text{ error}, \tag{4.3}$$

where $\boldsymbol{\beta}$, a p-vector of regression coefficients, and g, a smooth curve, are to be estimated. The model (4.3) is referred to as a *semiparametric* model, because the response Y is assumed to depend in a parametric (linear) fashion on some, but not all, of the explanatory variables. The variables $\mathbf{x}$ will be called the *linear* variables, while, for reasons that will become clear below, the variable t will be called the *splined* variable.

Such semiparametric models may seem rather a modest generalization of (4.1), but they are surprisingly useful. Very often in practice, the form of dependence of Y_i on most of the explanatory variables is known on grounds of theory or past experience, so there may be at most one or two candidates to be the splined variable, treated differently from the others. A particular context in which such a model arises very naturally is when a linear model is believed to be valid, except for possible inhomogeneity with respect to time. A semiparametric model (4.3) then allows the intercept to vary with time in a nonparametric way. An application to the analysis of agricultural field trials, in which $g(t)$ represents fertility at a spatial location t, will be considered in Section 4.6.

A rather less obvious application of the model (4.3) arises when the dependence of Y on a single explanatory variable t is supposed smooth except for discontinuities at prescribed values of t, or perhaps discontinuities in the first or second derivative. By introducing appropriate artificial variables $\mathbf{x}$, such features may be modelled using (4.3). Suppose, for example, that we wish to assume that the expected value of Y_i is a function $\mu(t_i)$ that is smooth and continuous except at 0. An artificial variable of the form $I[t_i > 0]$ will allow for a jump in μ at 0, while a jump in the qth derivative of μ would be allowed by including an artificial variable of the form $t_i^q I[t_i > 0]$. Note that if we allow a jump in μ we would normally include artificial variables corresponding to jumps in μ' and μ'' as well.

4.3 Penalized least squares for semiparametric models

If we wished to fit a semiparametric model (4.3) to data, we might attempt to estimate $\boldsymbol{\beta}$ and g by least squares, that is by minimizing

$$\sum_{i=1}^{n}\{Y_i - \mathbf{x}_i^T\boldsymbol{\beta} - g(t_i)\}^2.$$

However, in the absence of constraints on g, this approach will fail. Assume for the moment that the $\{t_i\}$ are distinct: whatever the value of $\boldsymbol{\beta}$, g can then be made sufficiently flexible to interpolate $g(t_i) = Y_i - \mathbf{x}_i^T\boldsymbol{\beta}$; therefore $\boldsymbol{\beta}$ is unidentifiable. In the spirit of Section 1.2.1, we circumvent this problem by instead choosing $\boldsymbol{\beta}$ and g to minimize the (weighted) penalized sum of squares

$$S_W(\boldsymbol{\beta}, g) = \sum_{i=1}^{n} w_i\{Y_i - \mathbf{x}_i^T\boldsymbol{\beta} - g(t_i)\}^2 + \alpha\int g''(t)^2 dt. \qquad (4.4)$$

The extra generality afforded by introducing weights $\{w_i\}$ will not be needed immediately, but is obtained at no real cost in complication.

4.3.1 Incidence matrices

In the discussion that follows, denote by $\mathbf{Y}$ the n-vector with i^{th} component Y_i, W the $n \times n$ diagonal matrix of weights w_i, and X the $n \times p$ model matrix whose i^{th} row is $\mathbf{x}_i^T$. In Chapter 2, we assumed that the $\{t_i\}$ were distinct and ordered: later, in Chapter 3, ties were handled in a fairly informal way. Now that we are dealing with multiple regression, in which it may be quite inconvenient to re-order all of the observed variables $\{Y_i, \mathbf{x}_i, t_i\}$, and the pattern of ties among the $\{t_i\}$ will not generally be related to that among $\{\mathbf{x}_i\}$, it is better to take a more formal approach. We will exclude the trivial case where all the t_i are identical, since this reduces to a parametric linear model!

Let the ordered, distinct, values among $t_1, t_2, \ldots, t_n$ be denoted by $s_1, s_2, \ldots, s_q$. The connection between $t_1, \ldots, t_n$ and $s_1, \ldots, s_q$ is captured by means of the $n \times q$ *incidence matrix* N, with entries $N_{ij} = 1$ if $t_i = s_j$, and 0 otherwise. It follows that $q \geq 2$ from the assumption that the t_i are not all identical.

4.3.2 Characterizing the minimum

Let $\mathbf{g}$ be the vector of values $a_j = g(s_j)$, so that $S_W(\boldsymbol{\beta}, g)$ can be written as

$$(\mathbf{Y} - X\boldsymbol{\beta} - N\mathbf{g})^T W(\mathbf{Y} - X\boldsymbol{\beta} - N\mathbf{g}) + \alpha\int g''(t)^2 dt.$$

Conceptually, the minimization of $S_W(\boldsymbol{\beta}, g)$ can be considered in two steps, first minimizing subject to $g(s_j) = a_j, j = 1, 2, \ldots, q$, and then minimizing the result over the choice of $\mathbf{g}$ and over $\boldsymbol{\beta}$.

The problem of minimizing $\int g''(t)^2 dt$ subject to g interpolating given points $g(s_j) = a_j$ where $s_1 < s_2 < \ldots < s_q$ was encountered in Section 2.2; the minimizing curve g is, of course, a natural cubic spline with knots $\{s_j\}$. Define matrices Q and R as in Section 2.1.2, but with $s_1, s_2, \ldots, s_q$ replacing $t_1, t_2, \ldots, t_n$, and define $K = QR^{-1}Q^T$. Theorem 2.1 then shows that the minimized value of $\int g''(t)^2 dt$ is $\mathbf{g}^T K \mathbf{g}$.

For this g, $S_W(\boldsymbol{\beta}, g)$ takes the value

$$(\mathbf{Y} - X\boldsymbol{\beta} - N\mathbf{g})^T W(\mathbf{Y} - X\boldsymbol{\beta} - N\mathbf{g}) + \alpha \mathbf{g}^T K \mathbf{g}. \tag{4.5}$$

By simple calculus or by completing the square it follows that (4.5) is minimized when $\boldsymbol{\beta}$ and $\mathbf{g}$ satisfy the block matrix equation:

$$\begin{bmatrix} X^T W X & X^T W N \\ N^T W X & N^T W N + \alpha K \end{bmatrix} \begin{pmatrix} \boldsymbol{\beta} \\ \mathbf{g} \end{pmatrix} = \begin{bmatrix} X^T \\ N^T \end{bmatrix} W\mathbf{Y}. \tag{4.6}$$

Notice that when the parametric part of the model involving X and $\boldsymbol{\beta}$ is omitted, these equations reduce to

$$(N^T W N + \alpha K)\mathbf{g} = N^T W \mathbf{Y}, \tag{4.7}$$

so that the smoother that has to be applied to $\mathbf{Y}$ to obtain the vector of fitted values $N\mathbf{g}$ is

$$S = N(N^T W N + \alpha K)^{-1} N^T W.$$

Readers should reassure themselves that this agrees with the solution to the smoothing spline problem with weights and possibly tied observations derived in Section 3.5. If, in addition, the t_i are distinct and already ordered, so that $N = I$, S further reduces to

$$S = (W + \alpha K)^{-1} W$$

as in equation (3.22).

4.3.3 *Uniqueness of the solution*

We need to do a little more work than was needed in Sections 2.3 and 3.5 to establish that our estimating equations have a unique solution. Once the uniqueness of $\mathbf{g}$ and $\boldsymbol{\beta}$ is established, the uniqueness of the curve g follows at once from the uniqueness of natural cubic spline interpolation.

We shall actually show that the block matrix appearing in (4.6) is positive-definite for all positive α. This immediately implies that (4.6) has a unique solution. Estimating equations of this form will recur at various points in later chapters, so it will be useful to prove this result in rather more generality than is needed here.

Theorem 4.1 *Suppose that W is an $n \times n$ symmetric positive-definite matrix, that X and N are any matrices of dimension $n \times p$ and $n \times q$ respectively, and that K is a $q \times q$ symmetric non-negative definite matrix.*

Let r be the rank of K, and T be any $q \times (q-r)$ matrix whose columns form a basis for the null space of K. If the block matrix $[X\ NT]$ is of full column rank, then

$$\begin{bmatrix} X^T WX & X^T WN \\ N^T WX & N^T WN + \alpha K \end{bmatrix}$$

is positive-definite for all positive α.

Proof. For any p-vector $\mathbf{u}$ and q-vector $\mathbf{v}$,

$$(\mathbf{u}^T \quad \mathbf{v}^T) \begin{bmatrix} X^T WX & X^T WN \\ N^T WX & N^T WN + \alpha K \end{bmatrix} \begin{pmatrix} \mathbf{u} \\ \mathbf{v} \end{pmatrix}$$

$$= (X\mathbf{u} + N\mathbf{v})^T W (X\mathbf{u} + N\mathbf{v}) + \alpha \mathbf{v}^T K \mathbf{v}. \tag{4.8}$$

From the assumptions on W and K, (4.8) is clearly non-negative. It can only be 0 if $X\mathbf{u} + N\mathbf{v} = 0$ and $\mathbf{v}$ is in the null space of K, so that $\mathbf{v} = T\boldsymbol{\delta}$ for some $\boldsymbol{\delta}$. This implies that $[X\ NT] \begin{pmatrix} \mathbf{u} \\ \boldsymbol{\delta} \end{pmatrix} = 0$, which forces $\boldsymbol{\delta} = 0, \mathbf{u} = 0$ under the assumption that $[X\ NT]$ has full rank. □

Note that the theorem makes no reference to W being diagonal, N being an incidence matrix, or K having the form derived from the penalty functional $\int g''(t)^2 dt$. Relaxation of each of these properties will be useful later. Note also that the column rank of $[X\ NT]$ will not be affected by the precise choice of T, as long as T satisfies the stated condition.

In the context of the present chapter, the hypothesis of Theorem 4.1 is quite easy to check. The only smooth curves for which $\int g''(t)^2 dt = 0$ are the first degree polynomials $g(t) = \delta_1 + \delta_2 t$, so T can be chosen to be the $q \times 2$ design matrix for linear regression. Therefore $[X\ NT]$ is of full column rank if and only if

(i) the columns of X are linearly independent, and

(ii) there is no linear combination $\mathbf{x}_i^T \boldsymbol{\beta}$ equal to a linear form $\delta_1 + \delta_2 t_i$ for all $i = 1, 2, \ldots, n$.

These are exactly the conditions for uniqueness of the least squares estimates in the fully parametric linear model with explanatory variables $(\mathbf{x}_i^T, 1, t_i)^T$.

4.3.4 Finding the solution in practice

Equation (4.6) forms a system of $p+q$ equations: this is typically very large, and it may not be convenient, or even practical, to solve this system directly. Fortunately, this is not necessary. One approach is to re-write (4.6) as the pair of simultaneous matrix equations

$$X^T WX\boldsymbol{\beta} = X^T W(\mathbf{Y} - N\mathbf{g}), \tag{4.9}$$

$$(N^T WN + \alpha K)\mathbf{g} = N^T W(\mathbf{Y} - X\boldsymbol{\beta}). \tag{4.10}$$

These are intuitively interpretable: (4.9) says that if g were known, we would subtract $(N\mathbf{g})_i = g(t_i)$ from Y_i, and estimate $\boldsymbol{\beta}$ by a weighted least squares regression of the differences. Conversely, if $\boldsymbol{\beta}$ were known, (4.10) tells us to fit a cubic smoothing spline to the differences $Y_i - \mathbf{x}_i^T\boldsymbol{\beta}$, as in (4.7).

This rather appealing interpretation can be taken further, because it turns out that we can use an alternation between equation (4.9) and (4.10), solving repeatedly for $\boldsymbol{\beta}$ and $\mathbf{g}$ respectively, to converge to the penalized least squares estimates. This procedure is sometimes known as *backfitting*, (see Breiman and Friedman (1985), Green, Jennison and Seheult (1985), and Buja, Hastie and Tibshirani (1989)) and in principle the iteration always converges, under the conditions in Theorem 4.1. Proof of this fact makes use of a simple result in matrix algebra, which we state here, as a second theorem.

Theorem 4.2 *Consider the linear equations*

$$\begin{bmatrix} A & B \\ B^T & C \end{bmatrix} \begin{pmatrix} \mathbf{u} \\ \mathbf{v} \end{pmatrix} = \begin{pmatrix} \mathbf{p} \\ \mathbf{q} \end{pmatrix}$$

and suppose that the block matrix is positive-definite, so that the equations have a unique solution. This solution is the limit of the iteration obtained by starting from any vector $\mathbf{u}^{(0)}$, *and repeatedly cycling between the two equations*

$$\mathbf{v}^{(n)} = C^{-1}(\mathbf{q} - B^T\mathbf{u}^{(n-1)}) \tag{4.11}$$

$$\mathbf{u}^{(n)} = A^{-1}(\mathbf{p} - B\mathbf{v}^{(n)}) \tag{4.12}$$

for $n = 1, 2, \ldots$.

Proof. Let $A = LL^T$, where L is non-singular. From (4.11) and (4.12),

$$\begin{aligned} L^T\mathbf{u}^{(n)} &= L^{-1}(\mathbf{p} - B\mathbf{v}^{(n)}) \\ &= L^{-1}(\mathbf{p} - BC^{-1}\mathbf{q} + BC^{-1}B^T\mathbf{u}^{(n-1)}) \\ &= L^{-1}(\mathbf{p} - BC^{-1}\mathbf{q}) + ML^T\mathbf{u}^{(n-1)} \end{aligned} \tag{4.13}$$

where $M = L^{-1}BC^{-1}B^T(L^T)^{-1}$. Now

$$\begin{vmatrix} \lambda A & B \\ B^T & C \end{vmatrix} = |C|\,|\lambda A - BC^{-1}B^T| = |C|\,|L|\,|\lambda I - M|\,|L^T|.$$

But clearly $\begin{bmatrix} \lambda A & B \\ B^T & C \end{bmatrix}$ is positive-definite for all $\lambda \geq 1$, so $|\lambda I - M|$ is non-zero for such λ. It therefore follows that all eigenvalues of M, which is non-negative definite, lie in the interval $[0, 1)$. The iteration (4.13) thus converges, and hence so does the backfitting cycle (4.11) and (4.12). □

In practice, the speed of convergence of this backfitting algorithm is governed by the size of α and how close $[X\ NT]$ is to being rank-deficient. For values of α of real interest, convergence is *usually* quite rapid, so that cycling between (4.9) and (4.10) can be terminated after perhaps 5 cycles. The equations involved in each cycle are cheap computationally and can be solved by standard methods. Equations (4.9) are simply weighted least squares equations, which take $O(n)$ time to solve, for fixed p, using routines to be found in any standard linear algebra library. As already noted, the equations (4.10) correspond to an application of spline smoothing, with smoothing parameter α, to the values $Y_i - \mathbf{x}_i^T\boldsymbol{\beta}$ at the points t_i with weights W. This can be carried out in $O(n)$ time once the data are ordered, for example by using the Reinsch algorithm allowing for weights and ties, by the obvious extension to Section 3.5.4. The equations can be written as

$$N\mathbf{g} = S(\mathbf{Y} - X\boldsymbol{\beta}), \tag{4.14}$$

where $S = N(N^TWN + \alpha K)^{-1}N^TW$ is the hat matrix of the spline smoothing operator for weights W and incidence matrix N.

4.3.5 A direct method

Although the backfitting approach set out in Section 4.3.4 yields an iterative scheme that always converges in theory, because the eigenvalues of the relevant matrix are all strictly less than one in absolute value, it can happen in practice that the largest eigenvalue is very near 1 and that the convergence is very slow. An example where this happens is given in Section 4.5 below.

An alternative approach allows the equations (4.6) to be solved in $O(n)$ time *without* any iteration. If we use (4.14) to eliminate $\mathbf{g}$ from (4.9), then we obtain the $p \times p$ linear system

$$X^TW(I - S)X\boldsymbol{\beta} = X^TW(I - S)\mathbf{Y} \tag{4.15}$$

for $\boldsymbol{\beta}$. These are generalized least squares normal equations, but with a non-diagonal weight matrix $W(I - S)$. Their solution might seem a computationally expensive problem, but once again the special structure of the smoothing operator S that is revealed by the Reinsch algorithm simplifies matters. Note that S can be applied to each column of X in $O(n)$ time, so that $X^T W(I-S)X$ can be constructed in $O(n)$ operations for fixed p. So can $X^T W(I-S)\mathbf{Y}$. The $p \times p$ linear system (4.15) can then be solved in $O(p^2)$ operations by a standard method, for example Cholesky decomposition. Finally, $N\mathbf{g}$ is found from (4.14), very efficiently since $S\mathbf{Y}$ and SX have already been constructed.

A disadvantage of this direct method of solution is that it is not apparently possible to combine it with orthogonal decomposition methods, to avoid having to form the $p \times p$ matrix $X^T W(I-S)X$. There is consequently some greater risk of rounding error.

4.4 Cross-validation for partial spline models

In Sections 3.2 and 3.5.3, we discussed the use of cross-validation to determine an appropriate value for the smoothing parameter automatically. The idea goes through, with appropriate modifications, in the partial spline case.

The cross-validation score still has the form

$$CV(\alpha) = \sum w_i \left(\frac{Y_i - \hat{Y}_i}{1 - A_{ii}} \right)^2 \tag{4.16}$$

where we now have

$$\hat{\mathbf{Y}} = A\mathbf{Y} = X\hat{\beta} + N\hat{\mathbf{g}}.$$

By (4.6), A is given by

$$[X\, N] \begin{bmatrix} X^T WX & X^T WN \\ N^T WX & N^T WX + \alpha K \end{bmatrix} \begin{bmatrix} X^T \\ N^T \end{bmatrix} W,$$

and, using (4.14) and (4.15), this can also be written as

$$S + (I-S)X\{X^T W(I-S)X\}^{-1} X^T W(I-S), \tag{4.17}$$

where

$$S = N(N^T WN + \alpha K)^{-1} N^T W. \tag{4.18}$$

The diagonal elements of S can be obtained via the Hutchison–de Hoog algorithm, exactly as in Section 3.2.2. As we have already said, finding $(I-S)X$ and $X^T W(I-S)$ are both $O(n)$ calculations. The matrix that must be inverted, $X^T W(I-S)X$, is $p \times p$ and can be found in $O(n)$ calculations. Finally, the matrix multiplications in (4.17) can be arranged to compute the *diagonal* elements of the result in $O(n)$ calculations, for fixed p.

The generalized cross-validation score is adapted from that in Section 3.3, and takes the form

$$GCV(\alpha) = \frac{\sum w_i(Y_i - \hat{Y}_i)^2}{(1 - n^{-1} \operatorname{tr} A)^2}$$

with A as defined in (4.17). The trace of A can be written

$$\operatorname{tr} A = \operatorname{tr} S + \operatorname{tr} [\{X^T W(I - S)X\}^{-1} X^T W(I - S)^2 X]$$

and of course this is also an $O(n)$ calculation.

As in Section 3.3.4, we define the equivalent degrees of freedom for noise by

$$EDF = \operatorname{tr} \{I - A\} = n - \operatorname{tr} A. \quad (4.19)$$

4.5 A marketing example

We will now give an example of the use of partial spline methods, and take the opportunity of comparing the results of the analysis with a more classical approach employing blocking.

Daniel and Wood (1980, pp. 142–145) discuss the analysis of data from a marketing price-volume study carried out in the petroleum distribution industry. Such a study would be used to examine the effect of price changes on the volume of sales, taking account of other relevant covariates. The conventional assumption is that the logarithm of the sales volume is linearly related to price. Here, the data consist of 124 observations on various days in the period February to August 1970. Data for only five days of the week are included (presumably for administrative reasons) and holiday dates were excluded. Thus the days of the year on which observations were taken are somewhat irregularly spaced in the period of interest.

The response variable Y is the log volume of sales of gasoline, and the two main explanatory variables of interest are x_1, the price in cents per gallon of gasoline, and x_2, the differential price to competition (that is, the amount by which the price charged by the company in question exceeds that of the average of its competitors). It is natural to expect that weekly and seasonal effects would be at work, so Daniel and Wood also included dummy variables to indicate the day of the week and the month in which each observation lay. They describe a complete analysis that includes an informal assessment of the influence of outlying values of x_1 and x_2.

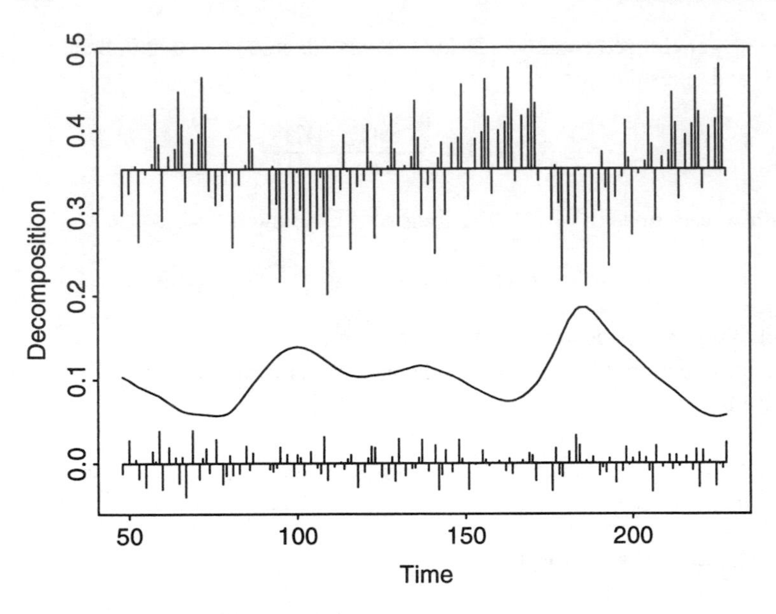

Figure 4.1. *Partial spline decomposition of the marketing data. Upper plot: parametric component of the fit; middle plot: dependence on splined variable; lower plot: residuals. All three plots are drawn to the same vertical scale, but the upper two plots are displaced upwards.*

4.5.1 A partial spline approach

In this situation, an attractive alternative is to model the dependence on time in a nonparametric fashion, so we now consider a semiparametric model in which the splined variable t represents the day of the year. Following Daniel and Wood (1980), we included weekly effects through dummy variables x_3, x_4, x_5, x_6, representing contrasts with the first day of the week. The method of Section 4.3.5 was used to fit the model, and the cross-validation score was found to be minimized at about $\alpha = 100$. For this degree of smoothing, the estimates of the regression coefficients β_1 and β_2 of primary interest were -0.0121 and -0.0060, with standard errors estimated as 0.0019 and 0.0025 respectively. As one might expect from the definition of the variables x_1 and x_2, there is a negative correlation, about -0.5, between the estimates of β_1 and β_2.

Aspects of this fit are displayed in Figure 4.1, in which a decomposition of the observed data is given. Three separate plots, all against the t_i, are given. At the top of the figure, the fitted *parametric* component $\mathbf{x}_i^T\hat{\beta}$ is shown. A close inspection of this part of the fit indicates the effect of the

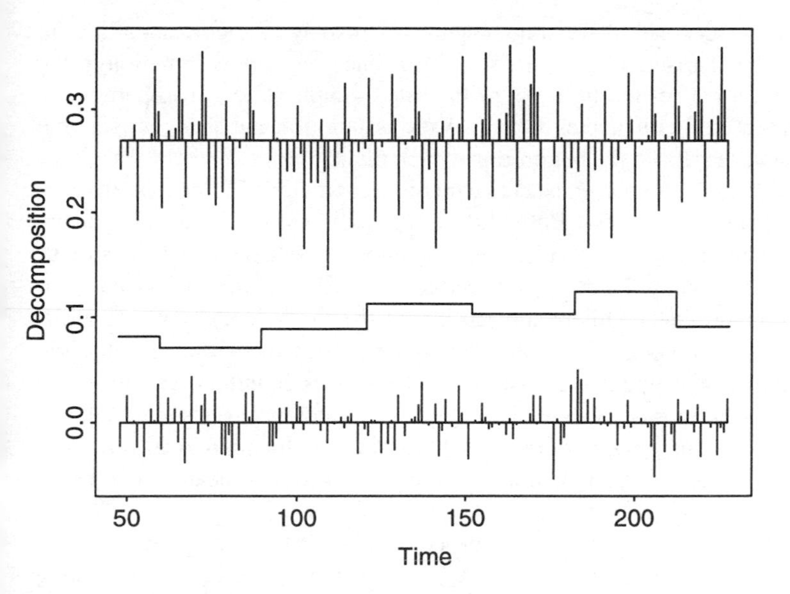

Figure 4.2. *Least squares decomposition of the marketing data, using calendar month as a blocking variable.*

day of the week on the sales. The longer term effects in this part of the fit are of course due to changes in the price variables x_1 and x_2. In the middle of the figure the nonparametric component $\hat{g}(t)$ is plotted as a continuous curve. Finally, at the bottom of the figure, the residuals $Y_i - \mathbf{x}_i^T\hat{\beta} - \hat{g}(t_i)$ are each plotted against t_i. It should be stressed that the origin on the y-axis applies to the residual plot only; the other two plots are drawn to the same vertical scale but are displaced vertically for clarity. The residual sum of squares was 0.0356; this has 99.3 equivalent degrees of freedom, as defined in (4.19). The difference $124 - 99.3 = 24.7$ in degrees of freedom for fitting 6 parametric covariates and the dependence on time, together with the appearance of the middle trace in Figure 4.1, indicate that the form of variation with time is quite complicated.

4.5.2 Comparison with a parametric model

By way of comparison, we also computed a fully parametric multiple regression fit to these data, treating calendar month as a factor, in place of the nonparametric trend used above. The estimates of β_1 and β_2 then became -0.0050 and -0.0110 respectively, with estimated standard errors of 0.0009 and 0.0020. The residual sum of squares was 0.0523 on 111 degrees of freedom. Figure 4.2 displays the corresponding fit, for comparison with Figure 4.1. This time, the stepped function in the central part of the figure is of the fitted month effects. The parametric model implicitly makes the unrealistic assumption that changes over time occur in steps at the beginning of each month.

In this example, it is the regression coefficients β_1 and β_2 that are of principal interest. Both the corresponding variables x_1 and x_2 are measured in cents, so it is fair to compare the estimated values of β_1 and β_2 directly. There is an important difference between the two sets of estimates. In the fully parametric model, the coefficient β_2 for differential price to competition is estimated as more than twice that for absolute price, suggesting that the volume of sales is more influenced by x_2 than by x_1. In contrast, with the partial spline, the comparison is reversed, and it is price (x_1) that is more influential. This contrast has clear relevance in the substantive application. Judged on a residual mean square basis, the partial spline model also fits the data better.

Another way of looking at this comparison is to consider the explanatory variables as being price (x_1) and average price of the competing brands ($x_1 - x_2$). Against these variables, the regression coefficients will be $\beta_1 + \beta_2$ and $-\beta_2$ respectively. The semiparametric model will therefore give as the part of the estimated model that depends on price variables

$$-0.0181 \times \text{ price } + 0.0060 \times \text{ average price of competition,}$$

while the fully parametric model with monthly blocking will give

$$-0.0160 \times \text{ price } + 0.0110 \times \text{ average price of competition.}$$

In both cases, temporal effects will also have to be added to give the full model. From this point of view, the effect of absolute price is estimated to be approximately the same whichever model is fitted; it is in evaluating the effect of the price of the competing brands that the models really differ.

A comparison between the curve $\hat{g}$ in Figure 4.1 and the corresponding stepped monthly effect curve in Figure 4.2 indicates why there is such a discrepancy between the two fits. The parametric curve peaks sharply around the beginning of July then falls off quite rapidly through the months of July and August. There is a similar, but much milder, effect near the beginning of April. While the monthly effects in the fully parametric model can be seen to correspond roughly to averages of the curve $\hat{g}$ over the relevant months, the blocking of the temporal effect into a function that is constant over months obscures the structure demonstrated in Figure 4.1, and does not allow such a good fit to the observed data.

We conclude this discussion with a remark about computing the partial spline fit. The iterative backfitting method (Section 4.3.4) effectively fails on these data, because it converges so slowly. In fact, an explicit calculation of the eigenvalues of the matrix M, defined in the proof of Theorem 4.2, shows them to be 0.99919, 0.44188, 0.07421, 0.01589, 0.01156 and 0.00467, when the smoothing parameter is set at 100. The closeness of the maximum eigenvalue to 1 explains the slow convergence, and the necessity to use the direct method set out in Section 4.3.5.

4.6 Application to agricultural field trials

A key idea in the design and analysis of experiments is the control of environmental variation so that comparison between treatments is not adversely affected by variability in experimental material; this is traditionally accomplished by appropriate blocking. In the context of agricultural field trials, blocks are physical areas of a field. The practical limitations of agricultural machinery, and other biological and statistical considerations, dictate a minimum plot size. Thus when the number of treatments is large, for example in cereal variety trials, the blocks needed for a randomized block design are enormous. Such blocks are demonstrably inhomogeneous with respect to important environmental factors such as soil nutrients, moisture, drainage and sunlight—in brief, 'fertility'—so there has been considerable recent interest in methods of analysis for field trials that account for variation on a finer spatial scale. Readers are referred to the paper by Wilkinson *et al.* (1983), and its accompanying discussion.

4.6.1 A discrete roughness penalty approach

A number of proposals have been made which effectively use a model of the form (4.3), in which $\mathbf{x}_i^T\boldsymbol{\beta}$ represent the treatment effects, t_i the spatial location of the ith plot, and $g(t_i)$ the fertility effect on this plot. In the case of cereal trials, plots are long and thin, so that it is usually considered sufficient to account for smooth variation in fertility in one dimension only. Among approaches of this nature are those of Besag and Kempton (1986), who take g to be a realization of a spatial autoregressive process, and Green, Jennison and Seheult (1983, 1985) who adopted a nonparametric approach termed *least squares smoothing* that is more in the spirit of this book.

Let X denote the design matrix for the treatment contrasts $\boldsymbol{\beta}$, the overall mean being subsumed into the fertility term. Then an appropriate model for Y_i, the yield on the ith plot, is

$$Y_i = (X\beta)_i + g(t_i) + \text{error}.$$

The function g is considered to be smooth, in the sense that its second differences $\{g(t_{i-1}) - 2g(t_i) + g(t_{i+1})\}$ are small. In the least squares smoothing procedure, β and g are estimated by minimizing the quadratic penalty function

$$\sum_{i=1}^{n}\{Y_i-(X\beta)_i-g(t_i)\}^2 + \alpha\sum\{g(t_{i-1})-2g(t_i)+g(t_{i+1})\}^2. \tag{4.20}$$

The plots in large cereals trials are typically laid out in several lines, so the second summation in (4.20) runs only over plots i for which $i-1, i, i+1$ are contiguous. The use of second differences in (4.20) is suggested by the discrete nature of the spatial scale in these experiments. It would be a modest change to replace the penalty term by an appropriate multiple of $\int g''^2$, to obtain the unweighted version of (4.4). Green *et al.* (1985) derive the estimating equations corresponding to (4.6), and suggest algorithms similar to those described in Sections 4.3.4 and 4.3.5.

From an algorithmic point of view, the discrete version of the roughness penalty offers a slight simplification. The matrix K is now banded: in fact if we choose the scale of t so that $t_i = i$, it is clear that $Q^T\mathbf{g}$ is the vector of second differences that appears in (4.20), so that $K = QQ^T$. It follows that the smoothing operator $S = (I + \alpha K)^{-1}$ is the inverse of a banded matrix, so may be applied to a vector in $O(n)$ time using a Cholesky decomposition, without need for the indirect approach of the Reinsch algorithm.

More traditional approaches to the analysis of field experiments are based on blocking. For the reasons mentioned earlier, a randomized complete blocks design and analysis is usually invalid, so that an incomplete blocks analysis is used, with Yates's recovery of intra-block information. This method is equivalent to a generalized least squares analysis based on an error model with two sources of random variation: plot effects and block effects. Accordingly, some recent 'neighbour' methods of analysis also use generalized least squares, but based instead on a covariance structure that allows inter-plot correlations to vary more continuously. The 'linear variance' model of Williams (1985) is an example.

The treatment estimating equations that arise from the least squares smoothing approach, analogous to (4.15), also have the form of generalized least squares equations. As is suggested by the form of the penalty function (4.20), the corresponding model for the covariances is generated by assuming that plot effects are uncorrelated with variance σ^2, while

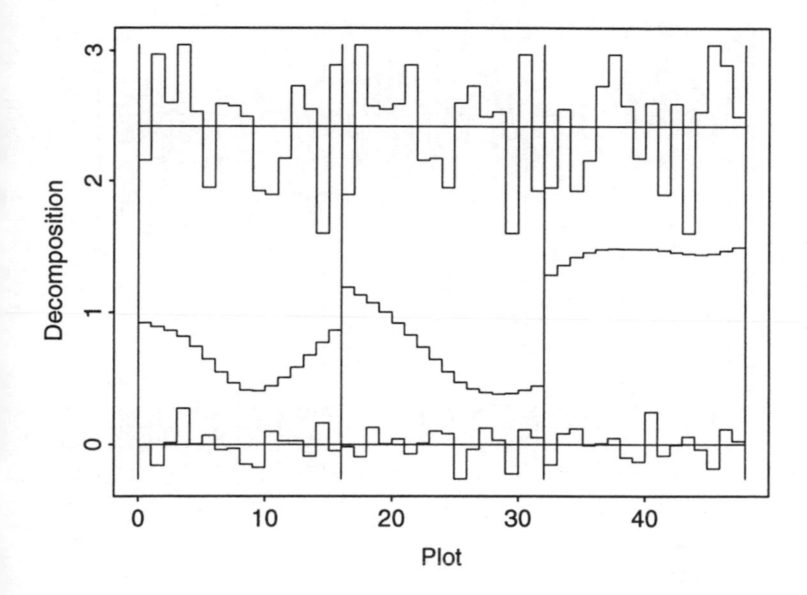

Figure 4.3. *Least squares smoothing decomposition of a small variety trial.*

the *second differences* of the fertility effects $g(t_i)$ are uncorrelated and have variances σ^2 / α. This is essentially the discrete analogue of the Bayesian justification for spline smoothing presented in Section 3.8.3, and closely parallels an argument proposed by Whittaker (1923) in an early application of roughness penalty methods in an actuarial context.

If the cubic spline roughness penalty $\int g''^2$ is used, it turns out that the corresponding model for the fertility effects supposes the second differences to have first-order moving average correlation structure, in which the lag-1 autocorrelation is $\frac{1}{4}$ (to see this, note the form of R when the t_i are equally spaced). This observation completes the connection between the neighbour methods that have been mentioned here: the linear variance model gives the same estimating equations as would a least squares smoothing method using *first* differences. The connections between smoothing methods and random effects linear models for spatial analysis of experiments are further explored by Green (1985).

4.6.2 *Two spring barley trials*

Here, we illustrate the least squares smoothing method applied to two cereal variety trials, using data kindly supplied by the Scottish Colleges of Agriculture. In Figure 4.3 is displayed the least squares smoothing

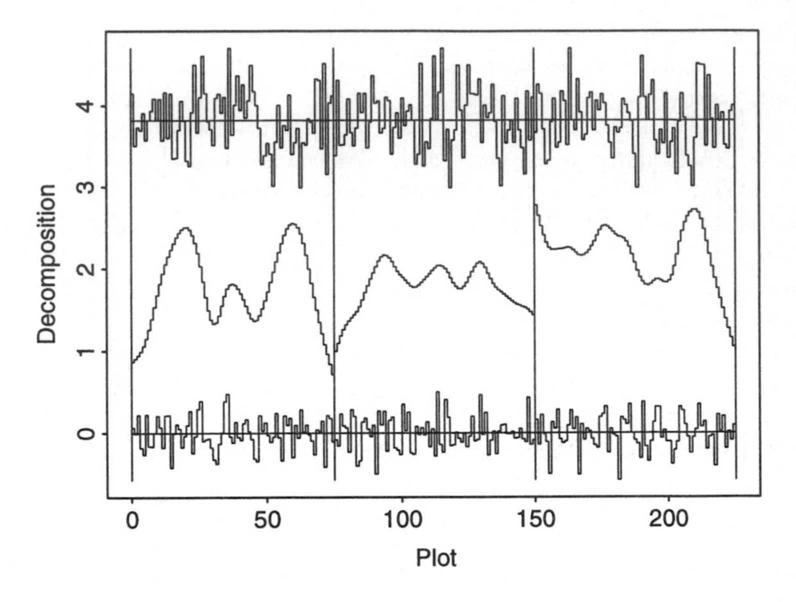

Figure 4.4. *Least squares smoothing decomposition of a larger variety trial.*

decomposition of data from a small trial of 16 varieties of spring barley, conducted in 1977. The varieties were sown in long narrow plots, approximately 2×20 metres, arranged linearly in 3 superblocks, each consisting of one complete replicate of the 16 varieties. The layout of the experiment was generated by a generalized lattice design, compatible with use of the standard incomplete blocks analysis. In our semiparametric analysis, smoothing is applied within each superblock only; this is consistent with the standard practice of including fixed replicate effects in the fitted linear model.

The decomposition in Figure 4.3 is constructed using a value of 10 for the smoothing parameter α; this is a compromise between the values suggested by a broad range of criteria discussed in Green *et al.* (1985) and Green (1985), which as applied to this data set give values between 5.2 and 16.9. Within this range there are only very small differences discernible between the different decompositions. The criteria, which include the two forms of cross-validation, two likelihood-based methods, and others based on classical variance-component methodology, are in accord with each other, largely because of the high degree of neighbour balance in the design of the experiment.

The layout of Figure 4.3 is similar to that used before, in Section 4.5.

The top trace shows the fitted variety effects, and the middle one the fitted fertility trends. In the lower trace can be seen the residuals, which appear unstructured. All three traces are plotted on the same vertical scale (but with origins displaced vertically), and it is evident that the variability in fertility identified by the analysis is very marked in two of the three superblocks. In fact, in the central superblock, the range of fertility is almost as large as the difference between the best and worst varieties. Except in the right hand superblock, the fertility pattern is seen to be more complicated than could be adequately approximated by either a subdivision into blocks or a trend linear in space.

In Figure 4.4, the analysis of a larger trial of 75 varieties is displayed; again, substantial and complicated patterns of fertility are revealed. These suggest that it is unduly optimistic to rely on randomization to eliminate bias in the standard analysis.

4.7 The relation between weather and electricity sales

In this section we discuss an ingenious application of semiparametric modelling that also illustrates another feature of roughness penalty methods. So far, it has almost invariably been assumed that the observations Y_i each depend on the unknown curve g at a single point t_i. In the example we now discuss, drawn from Engle, Granger, Rice and Weiss (1986), the dependence is of a more general form.

For many reasons, it is of interest to determine the relationship between temperature and the sales of electricity. In any kind of modelling of electricity demand—whether for economic reasons or assessing capacity requirements in the light of short-term weather forecasts—it is obviously important to take account of this relationship. In a country like the United States, where air conditioning is widely used, the consumption of electricity is high both on cold days and on hot days. It is common to assume, without any strong justification, that the demand for electricity is a V shaped function of temperature, with a minimum at 65°F. Engle *et al.* (1986) sought to relax this assumption by fitting an appropriate semiparametric model.

4.7.1 The observed data and the model assumed

Several data sets were considered. In each one, data for a time period covering several years were available. For each month $i = 1, \ldots, N$ the total sales in megawatt hours billed to consumers in that month are given. Let Y_i be the sales per customer in each month i. The sales in each month are the sum of a number of 'billing cycles'; each consumer is billed approximately on a monthly cycle, but on different days of the month. There are approximately 21 'billing cycles' relevant to each month. Thus if we are considering the March bills for example, the number of bills that include the consumption on February 22 will be the total number of customers on billing cycles ending in March that started on or before February 22. The exact start and end date of each cycle, and the number of customers in each cycle, are known. For each month i and each day j, let

$$m_{ij} = \text{number of bills presented in month } i \text{ that cover day } j$$

and let

$$M_i = \sum_j m_{ij} = \text{total number of bills presented in month } i.$$

The average temperature t_j on each day j in the period studied was given. In practice, the temperature range was divided into K small intervals and each recorded temperature was rounded to the midpoint of the interval in which it fell. Let the midpoints of the various intervals be $s_1, \ldots, s_K$.

In addition to temperature, a number of other variables were taken into account, all as linear variables. In order to account for seasonal effects not related to temperature (such as the length of daylight) 11 seasonal dummy variables were used, one for the difference between each calendar month and December. In addition there were variables corresponding to average household income and the unit price of electricity, both relative to the consumer price index.

The model then used for the data was

$$Y_i = \mathbf{x}_i^T \beta + l_i(g) + \text{ error}, \tag{4.21}$$

where $g(t)$ is a function of temperature that gives the effect of temperature on consumption on a daily basis, so that after taking other variables into account the expected use of electricity per consumer on a day with average temperature t is $g(t)$. This definition allows us to specify the quantities $l_i(g)$ as linear functionals of g, as we now set out.

The expected bill in month i for a particular consumer would, by assumption, be

$$\mathbf{x}_i^T \beta + \sum_{\text{days covered by bill}} g(t_j). \tag{4.22}$$

For each j the overall proportion of bills in month i that include exposure to the temperature on day j is $M_i^{-1} m_{ij}$, and so the contribution of the temperature that day to the overall average observed monthly bill would be $M_i^{-1} m_{ij} g(t_j)$. Summing over all possible days covered gives as the overall effect of temperature

$$l_i(g) = M_i^{-1} \sum_j m_{ij} g(t_j) = \sum_k L_{ik} g(s_k) \tag{4.23}$$

where the $N \times K$ matrix L is given by

$$L_{ik} = M_i^{-1} \sum_{\{j:t_j = s_k\}} m_{ij}.$$

The model (4.21) differs from the usual semiparametric model in that each observation Y_i depends not on an individual value of $g(t)$ but rather on the whole (or a large part) of the curve g through the linear functional $l_i(g)$.

4.7.2 *Estimating the temperature response*

A roughness penalty approach was used for the estimation of g, using a simple discretization approach, based on the points to which the temperatures were rounded. It can be considered as analogous to a fairly crude form of the basis functions approach to spline smoothing set out in Section 3.6 above.

Let $\mathbf{g}$ be the vector of values $g(s_k)$. A matrix K based in a natural way on second divided differences then gives

$$\mathbf{g}^T K \mathbf{g} \approx \int g''^2.$$

In assessing the goodness-of-fit of the model to the data, a weight matrix W was used to allow for an autocorrelated error structure in the data, as will be discussed further in Section 6.3. For further details of the construction of W and K, see Engle *et al.* (1986). For smoothing parameter α, it follows from (4.21) and (4.23) that the penalized weighted residual sum of squares is

$$S_W(\beta, g) = (\mathbf{Y} - X\beta - L\mathbf{g})^T W(\mathbf{Y} - X\beta - L\mathbf{g}) + \alpha \mathbf{g}^T K \mathbf{g} \tag{4.24}$$

which is easily minimized to give the estimates of the parameters β and the vector $\mathbf{g}$.

For a data set from Northeast Utilities in Hartford, Connecticut, USA, the estimated temperature response curve is given in Figure 4.5. The smoothing parameter α was chosen by generalized cross-validation. The values of the coefficients corresponding to the calendar month linear variables are given in Figure 4.6. Note that these are all relative to December, so the coefficient corresponding to December is automatically

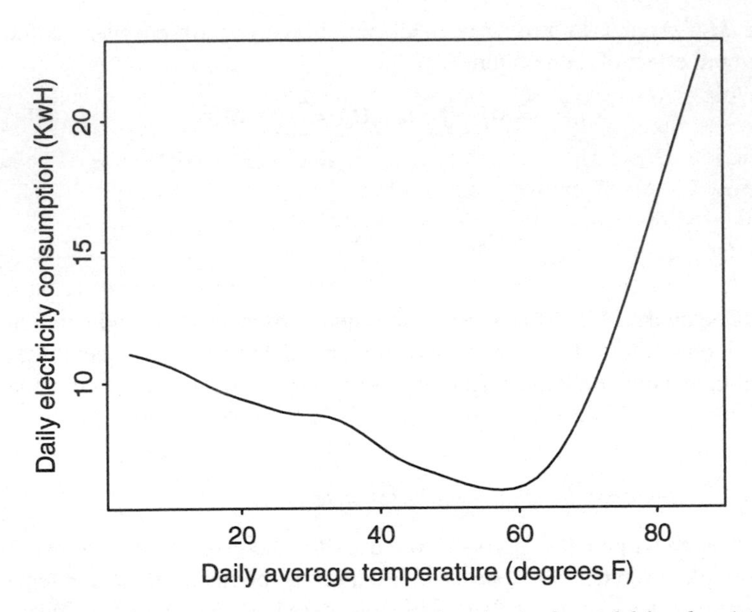

Figure 4.5. *Estimated temperature response function in the model for electricity consumption, for Northeast Utilities data. Results taken from Engle* et al. *(1986).*

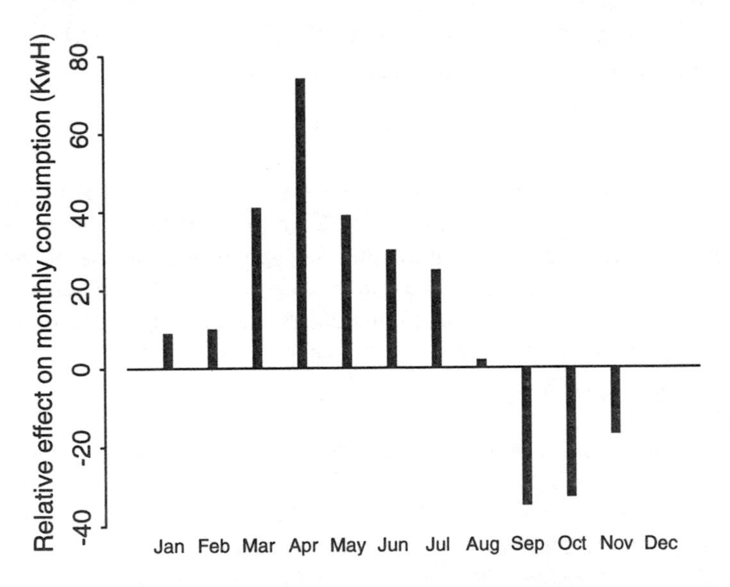

Figure 4.6. *Relative effect of calendar month in the model for electricity consumption, for Northeast Utilities data. Results taken from Engle* et al. *(1986).*

zero. It should be noted from (4.22) that a particular customer's predicted bill would include a single calendar variable coefficient but the sum of about thirty values of the temperature response curve, so the calendar month dummy variables in fact contribute far less to the overall prediction of electricity consumption. The analysis also gave coefficients showing a small positive effect due to average household income and a small negative effect due to price.

Overall, this analysis, in common with that of other data sets studied, confirms a pattern of increased consumption as the temperature moves away from 'comfortable' in either direction, with a sharper rate of increase for high temperatures. Reducing the temperature by air conditioning uses more electricity per degree than increasing it by heating. (Note that not all heating uses electricity.) The minimum consumption is at a temperature somewhat below 65°F. Perhaps somewhat surprisingly, the nonparametric response function retains the property of the V-shaped model of a linear increase in consumption at temperatures well away from the minimum point; the curve g is of course smooth near its minimum, and this is a much more realistic model than one with an abrupt change in behaviour at a particular point. Therefore the semiparametric model is preferable as an input into other modelling procedures. The pattern of monthly consumption shown in Figure 4.6 is of clear interest; all other things being equal, consumption is higher in the spring and lower in the autumn.

4.8 Additive models

One of our motivations in studying the semiparametric model (4.3) was as a first step in generalizing from the fully linear model (4.1) to the fully nonparametric regression (4.2). The semiparametric model retains the appealing property, important both in modelling and in computation, that the mean response is *additive* in the explanatory variables $x_1, x_2, \ldots, x_p, t$, even though it is nonlinear in the last of these.

In many ways, therefore, the natural next step is to consider the generalization to

$$Y_i = \sum_{j=1}^{p} g_j(t_{ij}) + \text{ error}, \tag{4.25}$$

where t_{ij} is the value of the jth variable for the ith observation. In this specification, the component functions could all be linear in parameters β, in which case this remains a linear model; if all but one of the functions is of this form, and the other unspecified but for the requirement of smoothness, this is a semiparametric model of the form that we have already discussed in detail. More generally, this provides a substantial new class of flexible regression models, retaining the interpretability provided by the additive structure, without enforcing a rigid parametric form of dependence on explanatory variables that may not be justified in the application under consideration.

Additive modelling has been pioneered and developed by Trevor Hastie and Robert Tibshirani, and their recent monograph (Hastie and Tibshirani, 1990) provides an excellent source for the subject. It is therefore unnecessary for us to go into details here, in spite of the practical importance of this class of models.

When an additive model is estimated using the penalized least squares approach, the functions $\{g_j\}$ are estimated to minimize

$$S_W(g_1, \dots, g_p) = \sum_{i=1}^{n} w_i\{Y_i - \sum_{j=1}^{p} g_j(t_{ij})\}^2 + \sum_{j=1}^{p} \alpha_j \int g_j''(t)^2 dt. \quad (4.26)$$

Here $\alpha_1, \alpha_2, \dots, \alpha_p$ are separate smoothing parameters for each of the component functions. Hybrid models in which some of the functions are specified as linear are specified in the obvious way.

If the same line of argument is followed as that in Section 4.3, simultaneous estimating equations of the form

$$(N_j^T W N_j + \alpha_j K_j)\mathbf{g}_j = N_j^T W(\mathbf{Y} - \sum_{k \neq j} N_k \mathbf{g}_k)$$

are obtained, analogous to (4.10). As soon as there is more than one nonparametric component, a direct, non-iterative computational approach similar to that described in Section 4.3.5 is not available, so the backfitting method has to be used, as in Section 4.3.4. Thus the weighted cubic spline smoother retains its important rôle as a computational tool even in this more complex class of models.

Hastie and Tibshirani (1990) discuss these computational issues in some detail, and also cover important inferential matters, such as the assignment of equivalent degrees of freedom to the fitted component curves.

The very nature of additive modelling perhaps discourages attention to possible interactions between explanatory variables. In ordinary linear modelling, inclusion of certain products of observed explanatory variables as terms in the regression model is often natural, but it is less clear how to make the analogous modification to an additive model without introducing an element of arbitrariness. This issue certainly justifies further research activity.

4.9 An alternative approach to partial spline fitting

4.9.1 *Speckman's algorithm*

In this section, we discuss an alternative approach to partial spline fitting, due to Speckman (1988). In order to explain the approach, consider, for the moment, a framework within which the explanatory variables $\mathbf{x}_i$ in the model

$$Y_i = \mathbf{x}_i^T \boldsymbol{\beta} + g(t_i) + \text{ error,} \tag{4.27}$$

can be thought of as themselves having a regression dependence on t_i, of the form

$$\mathbf{x}_i = \boldsymbol{\xi}(t_i) + \boldsymbol{\eta}_i, \tag{4.28}$$

where $\boldsymbol{\xi}$ is a vector of smooth functions of t, and the $\boldsymbol{\eta}_i$ are vectors of 'errors' η_{ij}. Define g_0 to be a function such that

$$g_0(t_i) = \boldsymbol{\xi}(t_i)^T \boldsymbol{\beta} + g(t_i), \tag{4.29}$$

so that, in a certain sense,

$$Y_i = g_0(t_i) + \text{ error.}$$

Taking the difference between (4.27) and (4.29) then gives

$$Y_i - g_0(t_i) = \{\mathbf{x}_i - \boldsymbol{\xi}(t_i)\}^T \boldsymbol{\beta} + \text{ error.} \tag{4.30}$$

This equation shows that the parameter vector $\boldsymbol{\beta}$ can be estimated by regressing the residuals of Y_i on those of the explanatory variables, in each case taking the residuals from the 'trend' of the relevant variable given t, without taking any other variables into account. There is no (explicit) dependence on the curve g.

For any given smoothing parameter α, let S be the hat matrix of spline smoothing with parameter α. As in Section 4.3.1, let X be the matrix with rows $\mathbf{x}_i^T$, and let Ξ be the matrix with rows $\boldsymbol{\xi}(t_i)^T$. The relation (4.30) then suggests the following procedure.

1. Use spline smoothing to estimate $\{g_0(t_i)\}$ and Ξ, yielding the estimates $S\mathbf{Y}$ and SX respectively.
2. Define $\tilde{\mathbf{Y}}$ and $\tilde{X}$ to be the residuals $(I - S)\mathbf{Y}$ and $(I - S)X$ respectively.
3. Estimate $\boldsymbol{\beta}$ by solving the regression equation corresponding to (4.30), $\tilde{\mathbf{Y}} = \tilde{X}\boldsymbol{\beta} +$ error, to yield

$$\hat{\boldsymbol{\beta}} = \{\tilde{X}^T \tilde{X}\}^{-1} \tilde{X}^T \tilde{\mathbf{Y}}. \tag{4.31}$$

4. Substitute the estimate $\hat{\boldsymbol{\beta}}$ back into (4.27) and obtain an estimate $\hat{g}$ by spline smoothing applied to the values $Y_i - \mathbf{x}_i^T \hat{\boldsymbol{\beta}}$.

This algorithm for estimating the various parameters can of course be applied whether or not the regression dependence (4.28) is plausible, although the logic of using the same smoother S for both $\mathbf{Y}$ and X in step 1 above does perhaps assume this. The method has the advantage that it does not involve any iteration.

Generalized cross-validation can be used, if desired, to choose the smoothing parameter. To obtain the hat matrix of the procedure, note that the vector of predicted values is given by

$$\begin{aligned} X\hat{\beta} + S(\mathbf{Y} - X\hat{\beta}) &= S\mathbf{Y} + (I - S)X\{\tilde{X}^T\tilde{X}\}^{-1}\tilde{X}^T(I - S)\mathbf{Y} \\ &= [S + \tilde{X}\{\tilde{X}^T\tilde{X}\}^{-1}\tilde{X}^T(I - S)]\mathbf{Y}, \end{aligned}$$

and hence the trace of the hat matrix is the same as that of the matrix $S + \{\tilde{X}^T\tilde{X}\}^{-1}\{\tilde{X}^T(I - S)\tilde{X}\}$, an $O(n)$ calculation. This trace can also be substituted into the formula (3.19) to give an estimate of the error variance. Standard errors for $\hat{\beta}$ can then be derived from (4.31) by standard multivariate calculations.

4.9.2 *Application: the marketing data*

As an example, we apply the technique to the marketing data discussed in Section 4.5 above. The value of the smoothing parameter chosen by generalized cross-validation is virtually identical to that obtained previously. With this value of the smoothing parameter, the estimates of β_1 and β_2 are -0.0148 and -0.0040. The estimated standard errors, 0.0019 and 0.0025 respectively, are the same as before to two significant figures, and the residual sum of squares, 0.0358 on 100.1 equivalent degrees of freedom, is virtually identical. The fitted model is not substantially different, but does, if anything suggest that the price of the competition is less important, because it leads to the part of the estimated model that depends on price variables being

$$-0.0188 \times \text{ price } + 0.0040 \times \text{ average price of competition.}$$

The decomposition corresponding to the estimates obtained in this way is shown in Figure 4.7; it can be seen that the estimate of g and the values fitted by the estimated parameters are virtually identical to those in Figure 4.1.

4.9.3 *Comparison with the penalized least squares method*

In this discussion, we refer to the estimate obtained by the method of Section 4.3 as the penalized least squares (PLS) estimates.

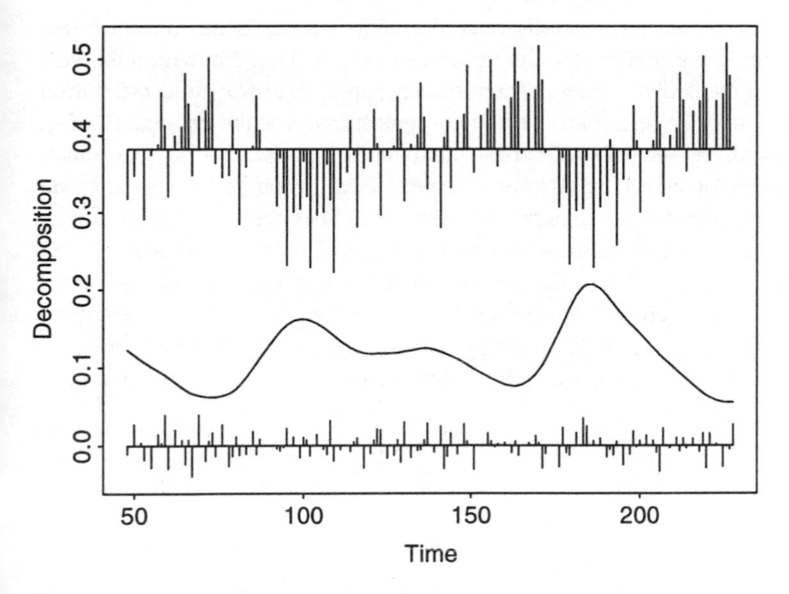

Figure 4.7. *Partial spline decomposition of the marketing data, estimated by Speckman's method, smoothing parameter chosen by generalized cross-validation.*

The original motivation for Speckman's algorithm as set out in Section 4.9.1 was a result of Rice (1986), who showed that, within a certain asymptotic framework, the PLS estimates of the parameters β could be susceptible to biases of the kind that are inevitable when estimating a curve. If the smoothing parameter is chosen to give good estimates of the curve g, then the biases in the estimates of β tend to zero, but at a rate that prevents the mean square error of the estimates from converging to zero at the usual 'parametric' $O(n^{-1})$ rate. Speckman (1988) showed that his method does not suffer from this difficulty; again under suitable conditions, if the smoothing parameter is chosen to give the best possible rate of consistency of $\hat{g}$ then the squared bias of $\hat{\beta}$ will be of smaller order than its variance, and the mean square error of $\hat{\beta}$ will tend to zero at the $O(n^{-1})$ rate.

It is, however, instructive to compare the formula (4.31) with the formula (4.15) for the PLS estimators. This comparison shows that the Speckman estimator for β (using smoothing matrix S) could be obtained in the same way as a PLS estimator, by replacing the smoothing matrix in the PLS method by $S_2 = I - (I - S)^2$. In the terminology of Tukey (1977), S_2 is the smoother obtained by 'twicing' S, and will correspond to a less severe smoothing operation than S itself. Thus in very broad terms Speckman's work can be interpreted as saying that within the PLS setup itself the amount of smoothing appropriate for good estimation of β may be less than the amount appropriate for the estimation of g. Indeed, Heckman (1986) proved that the PLS estimate $\hat{\beta}$ is consistent at parametric rates if small values of the smoothing parameter are used, and more general theoretical results along these lines are provided by Cuzick (1992). However the large-sample nature of all these results, and the various assumptions they depend on, suggest that more practical experience is required before it is possible to see whether the apparent theoretical advantage of the Speckman approach offsets the loss of interpretability and flexibility in departing from the penalized least squares paradigm.

CHAPTER 5

Generalized linear models

5.1 Introduction

All the models considered in the preceding chapters can be thought of as ways of extending ideas from linear regression. Various *nonlinear* or *non-normal* regression models have of course been studied on an individual basis for many years. However, only in 1972 did Nelder and Wedderburn provide a unified and accessible theoretical and computational framework for a class of such models, called *generalized linear models* (GLMs), which have been of enormous influence in statistics. In this chapter we set out ways in which roughness penalty methods can be applied in the broader context of generalized linear models.

5.1.1 Unifying regression models

We begin by unifying the models considered in the preceding chapters, in a manner that also makes them easy to generalize, by splitting the model for the observed data $\{Y_i\}$ into a *random component* and a *systematic component*. The basic idea is to introduce a vector of *predictors* θ_i, one for each observation. The random component of the model specifies the way in which the distribution of Y_i depends on θ_i, while the systematic component specifies the structure of the θ_i as a function of the available explanatory variables.

To deal first with the random component, note that in all of our development so far, we have been concerned with regression models with two key properties:

(a) Terms involving explanatory variables influence only the systematic part of the model.

(b) The models are appropriately fitted by estimating parameters using least squares, or penalized alternatives.

Although we have seldom mentioned probability models, the assumption in (b) is of course most strongly justified when the random component of the model follows a normal distribution: the maximum likelihood and least squares criteria then coincide. It is also, of course, the case that these assumptions hold for classical linear multiple or simple regression. Thus, in all the cases we have considered so far, we have at least tacitly assumed:

Random component:

$$Y_i \sim N(\theta_i, \sigma^2) \text{ for } i = 1, 2, \ldots, n. \tag{5.1}$$

Turning now to the systematic component, the three cases of classical multiple regression, univariate spline smoothing as considered in Chapters 2 and 3, and semiparametric models as considered in Chapter 4 can all be accommodated by suitable choice of model. We have the following possible assumptions:

Systematic component:

Parametric case (classical multiple regression):

$$\theta_i = \mathbf{x}_i^T \boldsymbol{\beta} \text{ for } i = 1, 2, \ldots, n \tag{5.2}$$

Nonparametric univariate case:

$$\theta_i = g(t_i) \text{ for } i = 1, 2, \ldots, n \tag{5.3}$$

Semiparametric case:

$$\theta_i = \mathbf{x}_i^T \boldsymbol{\beta} + g(t_i) \text{ for } i = 1, 2, \ldots, n. \tag{5.4}$$

These classes of models are of course applicable to a tremendous variety of problems across different disciplines. However it is obvious that these assumptions are restrictive, and there are many important regression problems in which a more general form of either the random component or the systematic component is appropriate.

5.1.2 Extending the model

The generalized linear models of Nelder and Wedderburn (1972) allow a far more general random component than (5.1) and extend (5.2) to allow θ_i to depend on the covariates through an arbitrary nonlinear function of the linear predictor $\mathbf{x}_i^T \boldsymbol{\beta}$. In the next section we review ordinary parametric GLMs. This is motivation for the main thread of this chapter, concerned with extending the scope of GLMs by introducing *semiparametric generalized linear models*, a synthesis of the ideas of GLMs and partial spline models.

The reasons presented earlier in this book for treating one or more explanatory variables in a nonparametric fashion do not apply only to linear models, and it is equally important to develop a semiparametric treatment of generalized linear models. As with linear models, we will not be concerned with relaxing the probabilistic assumptions governing the random component of the model, but with the linearity assumptions on the systematic component.

A special case that has been discussed in the literature in a somewhat piecemeal fashion is *nonparametric GLMs*. An early paper on this theme was Silverman (1978), which set out a form of nonparametric logistic regression. In its simplest form, linear logistic regression fits a model to zero-one observations Y_i at points t_i by assuming that

$$\operatorname{logit} P(Y_i = 1) = \beta_0 + \beta_1 t_i$$

for parameters β_0 and β_1 to be estimated. Nonparametric logistic regression replaces the straight line dependence by dependence on a smooth curve g, to yield

$$\operatorname{logit} P(Y_i = 1) = g(t_i).$$

This model can be seen to generalize both (5.1) and (5.3) by replacing the normal random component by

$$Y_i \sim \text{Bernoulli}(\theta_i)$$

and by introducing the logistic function to yield the nonlinear systematic component

$$\operatorname{logit} \theta_i = g(t_i).$$

As part of our unified treatment, we shall treat models of this kind as a special case of semiparametric GLMs, in Section 5.3 below, where we include a practical example involving binomial-logistic dependence.

After setting out and discussing semiparametric GLMs in full detail, we extend the model still further in Section 6.5 to cover more general regression models. Problems involving smoothing with respect to more than one variable are not discussed in detail but can be approached using the techniques described in Chapter 7 below. Finally in this chapter we shall return to models based only on first and second moment properties, rather than a fully specified probability model for the random component.

5.2 Generalized linear models

5.2.1 Exponential families

In this section, we follow Nelder and Wedderburn (1972) in relaxing each of the assumptions (5.1) and (5.2) while remaining within a parametric framework. We first relax the assumption of normality by supposing that the responses Y_i are drawn *independently* from a one-parameter exponential family of distributions, with density or probability function

$$p(y_i; \theta_i, \phi) = \exp\left(\frac{y_i\theta_i - b(\theta_i)}{\phi} + c(y_i, \phi)\right) \tag{5.5}$$

Here θ_i is the natural parameter of the exponential family, specific to Y_i, which will carry information from the explanatory variables, and ϕ is a nuisance or scale parameter common to all Y_i, analogous to σ^2 in (5.1). The specific form of the distribution is determined by the functions b and c. We will see some examples below.

The systematic component of the model is now defined by specifying the functional form of θ_i in terms of the explanatory variables for the ith response. This is conventionally done indirectly: recall that from the standard theory of exponential families

$$\mu_i = E(Y_i; \theta_i, \phi) = b'(\theta_i). \tag{5.6}$$

Parametric generalized linear models are constructed by assuming that there exists a *link function* G such that

$$G(\mu_i) = \mathbf{x}_i^T \beta. \tag{5.7}$$

The right hand side of this expression is called the *linear predictor*. When fitting such models to data, it is usual for the form of the distribution and the link function to be fixed, chosen on grounds of theory or experience, and then several candidate sets of explanatory variables entertained, by considering various forms of linear predictor.

Examples

1. If we set $b(\theta_i) = \frac{1}{2}\theta_i^2$, $\phi = \sigma^2$, and $c(y_i, \phi) = -\frac{1}{2}(y/\sigma)^2 - \log \sigma\sqrt{2\pi}$ in (5.5) and G to be the identity function in (5.7), then it is easy to see we recover the normal linear model (5.1) and (5.2).

2. With $b(\theta_i) = e^{\theta_i}$, ϕ set identically 1, $c(y_i, \phi) = -\log(y_i!)$ and G as the logarithmic function, we obtain a model for Poisson data with a multiplicative structure for explanatory variables: this is the log-linear model that forms the basis for contingency table analysis.

3. If we put $b(\theta_i) = m_i \log(1 + e^{\theta_i})$, ϕ identically 1, $c(y_i, \phi) = -\log \binom{m_i}{y_i}$ and G as a logit function:

$$\mathbf{x}_i^T \beta = \log \frac{\mu_i}{m_i - \mu_i}$$

or probit function:

$$\mathbf{x}_i^T \boldsymbol{\beta} = \Phi^{-1}\left(\frac{\mu_i}{m_i}\right)$$

(where Φ is the standard normal integral) then we obtain the models used in bioassay leading to logit and probit analysis, respectively. These are both examples where the functions b, c and G vary with i, as all involve the binomial denominator m_i. This should not cause confusion, however, and we leave the dependence of these functions on i tacit in the notation.

For each particular exponential family density (5.5), one particular choice of link function has particular significance mathematically, and to a lesser extent statistically. If G is the inverse of the function b', then θ_i coincides with the linear predictor $\mathbf{x}_i^T \boldsymbol{\beta}$; we call this particular G the *canonical* link function for the model. This choice slightly simplifies the algebra and the algorithms, and has the effect of making $X^T Y$ sufficient for $\boldsymbol{\beta}$, given ϕ.

Generalized linear models form quite a general class of probabilistic regression models. Some of the assumptions, the exponential family distributions, the independence and the linearity in combining the explanatory variables, are a little more restrictive than necessary and will be relaxed later, but this class is very important in practice, and is certainly adequate for introducing the ideas in this chapter. A discussion of extensions to the class of generalized linear models, taking a geometric perspective, and commenting on the implications for estimation algorithms, was given by Green (1989).

5.2.2 *Maximum likelihood estimation*

One advantage of the full probabilistic specification of the model is that a natural principle for fitting the models suggests itself: that of maximum likelihood estimation. We denote by $\hat{\boldsymbol{\beta}}$ the maximum likelihood estimate (m.l.e) of $\boldsymbol{\beta}$: that value obtained by maximizing the log-likelihood

$$\ell(\boldsymbol{\theta}, \phi) = \sum_{i=1}^{n} \left(\frac{Y_i \theta_i - b(\theta_i)}{\phi} + c(Y_i, \phi) \right)$$

derived from (5.5), with $\boldsymbol{\theta}$ linked to $\boldsymbol{\beta}$ through equations (5.6) and (5.7). Questions of existence and uniqueness have to be answered individually for different special cases: see Wedderburn (1976), for example.

There is nowadays seldom any reason to consider any other estimator of $\boldsymbol{\beta}$, but before the wide availability of computers made the numerical evaluation of the maximum likelihood estimates a triviality, a simple and popular alternative was to apply the link function G from (5.7) to the raw data Y_i and perform an ordinary linear regression of $G(Y_i)$ on $\mathbf{x}_i$ to obtain estimates of β by least squares: see for example Finney (1947) for this approach to probit analysis. The least squares estimator will generally have inferior performance compared with $\hat{\beta}$.

5.2.3 *Fisher scoring*

Nelder and Wedderburn (1972) proposed Fisher scoring as a general method for the numerical evaluation of $\hat{\beta}$ in generalized linear models. That is, given a trial estimate β, update to β^{new} given by

$$\beta^{new} = \beta + \left\{ E\left(-\frac{\partial^2 \ell}{\partial\beta\partial\beta^T} \right) \right\}^{-1} \frac{\partial \ell}{\partial \beta} \tag{5.8}$$

where both derivatives are evaluated at β, and the expectation is evaluated as if β were the true parameter value. Then β is replaced by β^{new} and the updating is repeated until convergence is obtained.

It turns out that for a GLM, these updating equations take the explicit form

$$\beta^{new} = (X^T W X)^{-1} X^T W \mathbf{z} \tag{5.9}$$

where $\mathbf{z}$ is the n-vector with i^{th} component

$$z_i = (Y_i - \mu_i)G'(\mu_i) + \mathbf{x}_i^T \beta,$$

and W the $n \times n$ diagonal matrix with

$$W_{ii} = \{G'(\mu_i)^2 b''(\theta_i)\}^{-1}.$$

Notice how the nuisance parameter ϕ has cancelled out: its value is not addressed during the iterative estimation of β.

To see the truth of (5.9), we need to evaluate the derivatives in (5.8); it is convenient to introduce the notation η_i for the linear predictor $\mathbf{x}_i^T\beta$. Now

$$\begin{aligned} \frac{\partial \ell}{\partial \eta_i} &= \frac{\partial \ell}{\partial \theta_i}\frac{d\theta_i}{d\eta_i} = \frac{\partial \ell}{\partial \theta_i} \div \left(\frac{d\eta_i}{d\mu_i}\frac{d\mu_i}{d\theta_i} \right) \\ &= \left(\frac{Y_i - \mu_i}{\phi} \right) \div \{G'(\mu_i) b''(\theta_i)\}. \end{aligned} \tag{5.10}$$

Clearly $\frac{\partial^2 \ell}{\partial \eta_i \eta_j} = 0$ if $i \neq j$, and while $\frac{\partial^2 \ell}{\partial \eta_i^2}$ involves higher derivatives of G and b, we see that its (negative) expectation does not:

$$\begin{aligned} E\left(-\frac{\partial^2 \ell}{\partial \eta_i^2} \right) &= \frac{d\mu_i}{d\eta_i} \div \{\phi G'(\mu_i) b''(\theta_i)\} \\ &= \{\phi G'(\mu_i)^2 b''(\theta_i)\}^{-1}. \end{aligned} \tag{5.11}$$

Let $\mathbf{z}^*$ be the n-vector with $z_i^* = (Y_i - \mu_i)G'(\mu_i)$. Then, in summary, we have from (5.10)

$$\phi \frac{\partial \ell}{\partial \eta} = W\mathbf{z}^* \tag{5.12}$$

and from (5.11)

$$\phi E\left(-\frac{\partial^2 \ell}{\partial \eta \partial \eta^T}\right) = W. \tag{5.13}$$

But by the chain rule, since $\eta = X\beta$, we have

$$\frac{\partial \ell}{\partial \beta} = X^T \frac{\partial \ell}{\partial \eta}$$

and

$$E\left(-\frac{\partial^2 \ell}{\partial \beta \partial \beta^T}\right) = X^T E\left(-\frac{\partial^2 \ell}{\partial \eta \partial \eta^T}\right) X.$$

Thus the Fisher scoring equations (5.8) become

$$\beta^{new} = \beta + (X^T W X)^{-1} X^T W \mathbf{z}^*$$

which we can more simply express in the required form (5.9).

5.2.4 *Iteratively reweighted least squares*

There are several other ways in which these calculations could have been arranged: the point of this representation is that it shows that each iteration of the Fisher scoring method for numerical evaluation of the m.l.e. is a weighted least squares regression of the *working response vector* $\mathbf{z}$ on the model matrix X with a *working weights matrix* W. This is therefore an example of an *iteratively reweighted least squares* calculation. Most statistical packages and subroutine libraries provide the basic routines needed for this computation. Methods based on orthogonal decompositions are usually to be preferred to those explicitly forming the $p \times p$ matrix $X^T WX$. In general, both $\mathbf{z}$ and W are functions of the current estimate β and need to be re-evaluated each iteration.

In the normal linear model of Example 1, it is readily seen that $\mathbf{z}$ is the same as $\mathbf{Y}$ and that W is the identity matrix, so no iteration is needed and (5.9) merely confirms that the maximum likelihood and least squares coincide in this case.

5.2.5 *Inference in GLMs*

Analysis of data based on a generalized linear model means more than merely estimating the regression coefficients. Among other things, we need measures of goodness-of-fit, a definition of residual, methods for model selection and other hypothesis testing, estimates of standard errors of estimators, and confidence intervals. Here, we briefly review some of the basic methods.

The *scaled deviance* D^* is defined as the log-likelihood-ratio statistic $D^* = 2[\ell_{max} - \ell\{\theta(\hat{\beta})\}]$ where ℓ_{max} is the maximized log-likelihood for the *saturated model* allowing one parameter for each observation. If the model includes a nuisance parameter ϕ, it is usually easier to work with the unscaled version, the *deviance D* given by

$$\begin{aligned} D = \phi D^* &= 2\sum_{i=1}^{n}\{Y_i(\tilde{\theta}_i - \hat{\theta}_i) - b(\tilde{\theta}_i) + b(\hat{\theta}_i)\} \\ &= \sum_{i=1}^{n} d_i, \qquad \text{say.} \end{aligned}$$

Here, $\tilde{\theta}_i$ denotes the solution to $b'(\tilde{\theta}_i) = Y_i$, that is the value of θ_i maximizing the likelihood for the ith observation alone. The contribution d_i to the deviance from the ith observation is sometimes called the *deviance increment.*

The deviance D is a measure of the closeness of the fit of the model to the data, and can be interpreted very much like the residual sum of squares in a linear model, which is indeed what it reduces to, for the normal distribution/identity link GLM. An alternative measure of the quality of fit of the model to the data is provided by the Pearson chi-squared statistic

$$\chi^2 = \phi \sum_{i=1}^{n} \frac{\{Y_i - E(Y_i)\}^2}{\text{var}(Y_i)} = \sum_{i=1}^{n} \frac{(Y_i - \mu_i)^2}{b''(\theta_i)} \tag{5.14}$$

where $E(Y_i)$ and $\text{var}(Y_i)$ are evaluated at the maximum likelihood estimates $\hat{\beta}$. The relative merits of χ^2 and D are discussed in depth by McCullagh and Nelder (1989, pp. 33–36 and 118–122). They point out that the commonly-held assumption that each has approximately the χ^2_{n-p} distribution, after scaling by an estimate of ϕ if necessary, can be seriously misleading.

For model selection, a hierarchy of models can be investigated using deviances in a similar manner to the sums of squares in analysis of variance: this procedure is known as the *analysis of deviance*. This relies on asymptotic distribution theory; it can generally be assumed that *differences* between (scaled) deviances follow χ^2 distributions with the appropriate degrees of freedom, to a reasonable approximation. The extent to which such results hold as good approximations with finite sample sizes is still being assessed. For recent results in this direction see

Cordeiro (1985) and Barndorff-Nielsen and Blæsild (1986): the analysis can be alarmingly complicated.

For a *local* assessment of goodness-of-fit, to be used in checking both model adequacy and data adequacy (e.g. absence of outliers), it is useful to be able to examine residuals. How should these be defined? In models for independent responses, we would like to assign a residual z_i to each observation, that measures the discrepancy between Y_i and its value predicted by the fitted model, preferably on a standardized scale. Experience with residuals in multiple linear regression (see, for example, Cook and Weisberg, 1982) tells us that raw residuals $(Y_i - \mu_i)$ have unequal variances and are correlated, as a consequence of fitting the model. The first of these difficulties can be handled quite easily by dividing each raw residual by an estimate of its standard deviation to give a *standardized residual*, but the second is inevitable if residuals are to retain a one-to-one correspondence with the responses.

In generalized linear models there are two additional difficulties: firstly the model variances depend on the expectations, and secondly it is not obvious that data and fitted value should be compared on the original scale of the responses. These considerations lead to two commonly used definitions of residual:

$$z_i = \frac{Y_i - \mu_i}{\sqrt{b''(\theta_i)}}$$

and

$$z_i = \mathrm{sign}(Y_i - \mu_i)\sqrt{d_i}$$

where d_i is the deviance increment defined above. These are known as Pearson and Deviance residuals respectively. Their respective merits are discussed by Green (1984) and McCullagh and Nelder (1989, pp. 37–40). Each definition may further be standardized to adjust approximately for additional inequalities in variance introduced by estimation.

The asymptotic variance matrix of the maximum likelihood estimator $\hat{\beta}$ is

$$\left\{E\left(-\frac{\partial^2 \ell}{\partial \eta \partial \eta^T}\right)\right\}^{-1} = \phi(X^T W X)^{-1}.$$

Of course, in using this variance, W will be evaluated at the maximum likelihood estimate $\hat{\beta}$, and ϕ will be replaced by an estimate. McCullagh and Nelder (1989, p. 295) advocate using the Pearson χ^2 statistic to estimate ϕ (at least in the context of the gamma distribution):

$$\hat{\phi} = \frac{\chi^2}{(n-p)},$$

where χ^2 is as defined in equation (5.14).

5.3 A first look at nonparametric GLMs

5.3.1 *Relaxing parametric assumptions*

Before starting a more formal and comprehensive development of the ideas of nonparametric and semiparametric generalized linear models, it is helpful to present the details worked out in a rather straightforward special case, corresponding to the nonparametric univariate case (5.3). Suppose, therefore, that we observe responses $Y_i, i = 1, 2, \ldots, n$, drawn independently from the exponential family density (5.5), where now, instead of the parametric assumptions

$$G\{b'(\theta_i)\} = \mathbf{x}_i^T \beta$$

involving the link function and linear predictor, we suppose simply that

$$\theta_i = g(t_i)$$

where t_i are observed values of a one-dimensional explanatory variable, and g is an unknown but smooth function. We will assume in this section, as in Chapter 2, that the t_i are distinct and ordered, and lie within an interval $[a, b]$: $a < t_1 < t_2 < \ldots < t_n < b$. This defines a *nonparametric generalized linear model*. By virtue of the fact that there is no additional level of indirection between values of θ and the smooth function, we are implicitly adopting the link function that is canonical for the assumed density; this point will be elaborated later.

5.3.2 *Penalizing the log-likelihood*

If we attempt to maximize the log-likelihood

$$\ell(g, \phi) = \sum_{i=1}^{n} \left(\frac{Y_i g(t_i) - b\{g(t_i)\}}{\phi} + c(Y_i, \phi) \right) \tag{5.15}$$

over all smooth functions g, the result is useless. It is always possible to choose g sufficiently complicated that it interpolates the data, in the sense now that the fitted values agree with the observed responses: $Y_i = b'\{g(t_i)\}$. This is directly analogous with the similar point made about curve fitting in previous chapters, and and we adopt an analogous strategy to deal with it. Instead of maximizing the log-likelihood $\ell(g, \phi)$ alone, we choose $\hat{g} \in S_2[a, b]$ to maximize the *penalized log-likelihood*

$$\ell(g, \phi) - \tfrac{1}{2}\lambda \int g''(t)^2 dt \tag{5.16}$$

where λ is a smoothing parameter. The function $\hat{g}$ is called the *maximum penalized likelihood estimate* (MPLE).

It will generally be convenient to set $\alpha = \lambda\phi$ and to consider α as being the smoothing parameter. Multiplying (5.16) by ϕ shows that the problem of maximizing the penalized log-likelihood is precisely equivalent to that of maximizing

$$\sum_{i=1}^{n}[Y_i g(t_i) - b\{g(t_i)\}] - \tfrac{1}{2}\alpha \int g''(t)^2 dt, \tag{5.17}$$

eliminating the dependence on ϕ altogether. We shall define

$$\ell_1(g) = \sum_{i=1}^{n}[Y_i g(t_i) - b\{g(t_i)\}]$$

so that $\ell_1(g)$ differs from $\phi\ell(g,\phi)$, if at all, only by a term independent of g.

Motivation for the factor $\frac{1}{2}$ in (5.16) follows by noting that in the normal linear case,

$$\ell(g,\phi) = -\frac{1}{2\sigma^2}\sum\{Y_i - g(t_i)\}^2.$$

Since $\phi = \sigma^2$, $\phi\ell(g,\phi) = -\frac{1}{2}\sum\{Y_i - g(t_i)\}^2$, and so the maximization of (5.17) is equivalent to the minimization of

$$\sum\{Y_i - g(t_i)\}^2 + \alpha\int g''(t)^2 dt,$$

which is exactly the penalized least squares criterion used in Chapter 2.

Concentrating on the penalized log-likelihood (5.16) allows us to balance fidelity to the data (high values of the log-likelihood) with smoothness of the fitted curve g (low values of the roughness penalty). As in Section 3.8.3, there is a Bayesian argument leading to this criterion: the MPLE of g is the posterior mode given the data where ϕ is fixed and g has prior density proportional to $\exp(-\frac{1}{2}\lambda\int g''^2)$ over some suitable space of smooth functions g.

5.3.3 Finding the solution by Fisher scoring

By an argument now familiar, maximization of (5.17) follows a two-step principle. The first term in (5.17) depends on g only through the values $g(t_i)$, so the whole expression is maximized for fixed vector $\mathbf{g}$, with i^{th} component $g_i = g(t_i)$, by the natural cubic spline with knots $\{t_i\}$ interpolating $\mathbf{g}$. We know the resulting value for the second term is $-\frac{1}{2}\alpha\mathbf{g}^T K\mathbf{g}$, so we have only to maximize

$$\ell_1(\mathbf{g}) - \tfrac{1}{2}\alpha\mathbf{g}^T K\mathbf{g} \tag{5.18}$$

over choice of the n-vector $\mathbf{g}$. Here we have committed a harmless abuse of notation in writing $\ell_1(\mathbf{g})$ for the corresponding (scaled) log-likelihood term

$$\sum \{Y_i g_i - b(g_i)\}. \tag{5.19}$$

In general, the maximization of (5.18) will be a nonlinear optimization problem, and, as in the previous section, we use the approach of Fisher scoring to solve it.

Theorem 5.1 *The Fisher scoring algorithm for maximizing the penalized log-likelihood (5.18) with respect to* $\mathbf{g}$ *is given by*

$$\mathbf{g}^{new} = (W + \alpha K)^{-1} W \mathbf{z} \tag{5.20}$$

where

$$z_i = g_i + \frac{Y_i - b'(g_i)}{b''(g_i)},$$

and W is a diagonal matrix with

$$W_{ii} = b''(g_i).$$

Proof. We need the first and expected second derivatives of (5.18) with respect to $\mathbf{g}$. Immediately from (5.18) and (5.19), we find the ith component of the first derivative to be

$$\{Y_i - b'(g_i)\} - \alpha(K\mathbf{g})_i = \{(\mathbf{Y} - \boldsymbol{\mu}) - \alpha K \mathbf{g}\}_i,$$

where we have written $\boldsymbol{\mu}$ for the vector with $\mu_i = b'(g_i)$, while the ith diagonal element of the second derivative is

$$-\, b''(g_i) - \alpha K_{ii}.$$

The off-diagonal elements vanish, as in Section 5.2.3. Note that the second derivatives do not involve the responses $\{Y_i\}$, effectively because we are using the canonical link function, so that Newton–Raphson and Fisher scoring coincide in this problem. The algorithm becomes

$$\begin{aligned}
\mathbf{g}^{new} &= \mathbf{g} - \left\{ E\left(\frac{\partial^2 \ell_1}{\partial \mathbf{g} \partial \mathbf{g}^T}\right) - \alpha K \right\}^{-1} \left(\frac{\partial \ell_1}{\partial \mathbf{g}} - \alpha K \mathbf{g}\right) \\
&= \mathbf{g} + (W + \alpha K)^{-1} \{(\mathbf{Y} - \boldsymbol{\mu}) - \alpha K \mathbf{g}\} \\
&= (W + \alpha K)^{-1} \{W\mathbf{g} + \alpha K \mathbf{g} + (\mathbf{Y} - \boldsymbol{\mu}) - \alpha K \mathbf{g}\} \\
&= (W + \alpha K)^{-1} W \{\mathbf{g} + W^{-1}(\mathbf{Y} - \boldsymbol{\mu})\}
\end{aligned}$$

and this is the same as (5.20). □

Table 5.1. *A mortality table (from London, 1985, Table C-1). For each age x, the population size is n_x and the number of deaths is d_x*

x	n_x	d_x	x	n_x	d_x	x	n_x	d_x
55	84	1	72	20116	480	89	510	97
56	418	2	73	18876	537	90	430	93
57	1066	10	74	17461	566	91	362	75
58	2483	21	75	15012	581	92	291	84
59	3721	35	76	11871	464	93	232	31
60	5460	62	77	10002	461	94	196	75
61	6231	50	78	8949	433	95	147	29
62	8061	55	79	7751	515	96	100	25
63	9487	88	80	6140	374	97	161	20
64	10770	132	81	4718	348	98	11	5
65	24267	267	82	3791	304	99	10	3
66	26791	300	83	2806	249	100	8	2
67	29174	432	84	2240	167	101	5	0
68	28476	491	85	1715	192	102	4	2
69	25840	422	86	1388	171	103	2	0
70	23916	475	87	898	126	104	2	1
71	21412	413	88	578	86			

5.3.4 Application: estimating actuarial death rates

One of the oldest applications of the roughness penalty idea (see Whittaker, 1923) is in the smoothing of tables for actuarial purposes. It is interesting to note that Whittaker's paper uses a roughness penalty proportional to the sum of squares of third differences of the observations, and a weighted residual sum of squares as the measure of fit to the data. A Bayesian justification is used for the method. In this section, we give a somewhat different example of the application of penalized likelihood estimation to data that arise in the actuarial context.

A key component of actuarial work is the analysis of mortality tables, such as the one shown in Table 5.1. This table gives, for a particular population of retired American white females, the age structure of the population and the annualized number of deaths in each age group. These data have also been considered by London (1985) and by Ramsay (1993).

If we let the number of individuals in the population of age x be n_x and the number of deaths among these individuals be d_x, then a crude estimate of the mortality rate at age x is, of course, d_x/n_x. (We follow actuarial convention and use x as an integer suffix denoting age.) A plot of the crude rates derived from Table 5.1 is given in Figure 5.1. Actuaries

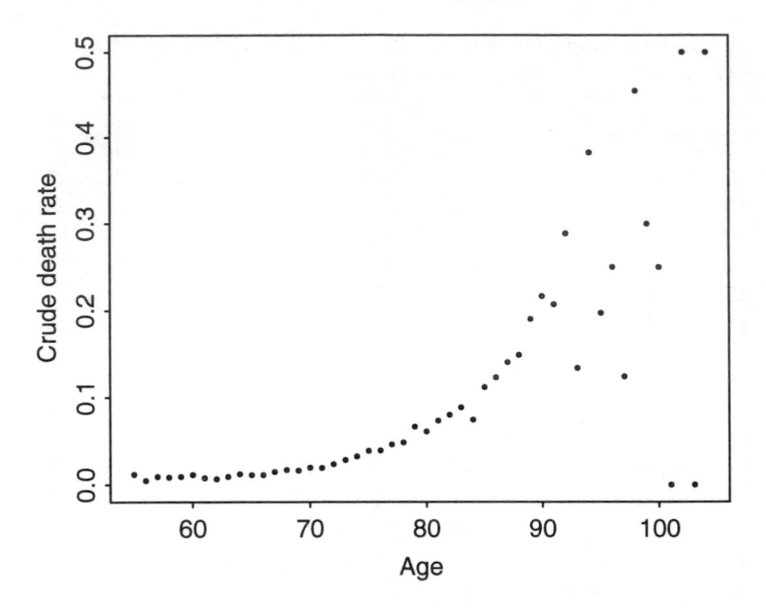

Figure 5.1. *Crude death rates.*

always assume that the mortality rate varies smoothly with age, and so these crude rates are smoothed in some way to obtain estimates of underlying 'true' death rates. In the actuarial literature, this smoothing process is called *graduation*.

A reasonable model for the observed data is to work conditionally on the values n_x and to assume that the d_x are drawn from independent binomial distributions with parameters n_x and $q(x)$, where $q(t)$ is a smooth curve giving the death rate at age t. A little care is required in the interpretation of $q(t)$ for non-integer ages t; this depends on the precise method of construction of the mortality table.

Clearly it is necessary to work in terms of a transform of q. We shall define a curve g by

$$g(t) = \text{logit}\, q(t) = \log \frac{q(t)}{1 - q(t)}.$$

By standard manipulations, the log partial likelihood of the curve g conditional on the values n_x is then, up to a constant,

$$\ell(g) = \sum_{x=55}^{104} [d_x g(x) - n_x \log\{1 + \exp g(x)\}].$$

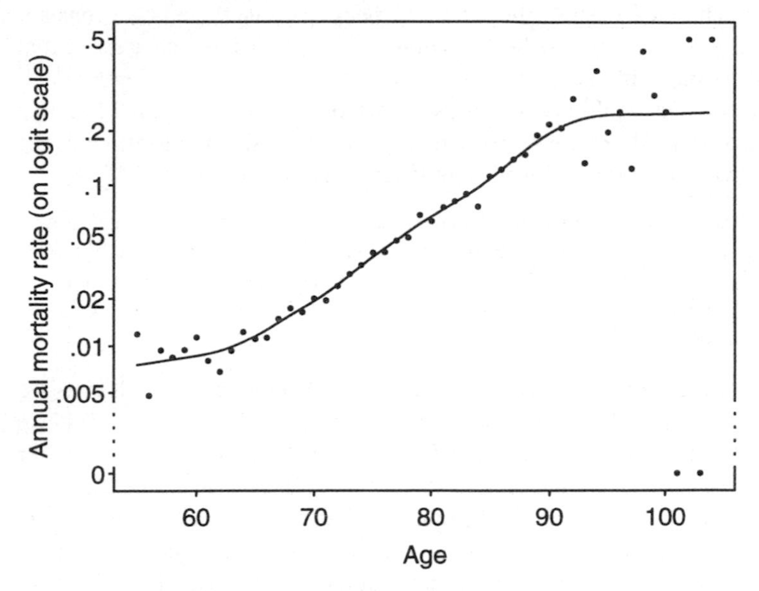

Figure 5.2. *Penalized likelihood estimate (—) of the death rate, and the crude death rates (•), both shown on a logit scale.*

The penalized likelihood estimate of g is then the curve that maximizes $\ell(g) - \frac{1}{2}\lambda \int g''^2$. This is easily found using the algorithm of Theorem 5.1. The binomial distribution is of the form (5.15) with b given by $b(\theta) = n_x \log(1 + e^\theta)$, and $\phi = 1$. Differentiation of this twice shows that the vector $\mathbf{z}$ has components

$$g(x) + \frac{d_x - n_x q(x)}{n_x q(x)\{1 - q(x)\}},$$

while the matrix W has diagonal elements $n_x q(x)\{1 - q(x)\}$. Theorem 5.1 shows that the MPLE of g is obtained by repeatedly calculating the cubic spline smoother of the working response vector $\mathbf{z}$ against age x using weights W, followed by updating $q(x)$ to $\exp\{g(x)\}/\{1 + \exp g(x)\}$. For a value of λ chosen by GCV, in the way described in Section 5.4.3, this yields the curve shown in Figure 5.2. This figure also shows the crude death rates plotted on a logistic scale; note that the two (very small) cohorts of ages 101 and 103 each contain zero deaths, so the logit of their crude death rates is $-\infty$. It can be seen that the estimated death rate is approximately linear on the logistic scale between the ages of about 65 and 92, but is flatter both for ages below 65 and those above 92. On the basis of these data (alone), it appears that once a member of this population has reached the age of 92, her death rate would be a constant of about 0.25, and so her remaining life would have an exponential distribution with mean about 4 years.

Actuaries would not, of course, work directly from a graph like this one but would use the numerical values of the estimated death rates for further calculations. Indeed, Whittaker (1923) wrote

> Workers in experimental science generally ... [plot the data] ... and draw a freehand curve as nearly as possible through them. This somewhat arbitrary method is insufficient for the needs of Actuarial Science, and a large number of "graduation formulae" are to be found in the journals of the Actuarial Societies.

It is also customary for actuaries to assume that the 'true' death rates are even smoother than those estimated in Figure 5.2. If, in order to achieve greater smoothness, a somewhat larger value of the smoothing parameter is used, then the overall pattern of the curve remains much the same.

The great advantage of the penalized likelihood approach to these data is that it takes account automatically of the fact that the variability of crude death rates is very different in different parts of the age range, because of the variation both in the size of the cohort and in the underlying death rate.

5.4 Semiparametric generalized linear models

In this section we now move to a more general situation, the GLM analogue of the semiparametric models discussed in Chapter 4. Following the approach of Green and Yandell (1985), we consider replacing the linear predictor $G(\mu_i) = \mathbf{x}_i^T \boldsymbol{\beta}$ from (5.7) by

$$G(\mu_i) = \mathbf{x}_i^T \boldsymbol{\beta} + g(t_i) \tag{5.21}$$

where $\mathbf{x}_i$ and t_i, both possibly vector-valued, are two sets of explanatory variables for the ith response, and the p-vector $\boldsymbol{\beta}$ and function g are to be estimated.

In some applications, such a model might arise when a parametric model is held to be appropriate on grounds of theory or experience, but there are doubts about the homogeneity of the model between situations distributed in time or space: the simple additive predictor (5.21) allows the 'intercept' term in, say, a logistic regression to vary in a nonparametric fashion. In other circumstances, the dependence of the distribution of Y_i on t_i is of more central interest, and indeed the parametric component $\mathbf{x}_i^T \boldsymbol{\beta}$ of the predictor may not be present.

5.4.1 *Maximum penalized likelihood estimation*

As with linear models, when there is an infinite-dimensional parameter involved, pure maximum likelihood is not an appropriate principle for model-fitting: but we will be interested in fitting a *smooth* function g, so consider maximizing a penalized likelihood. In this chapter we will only consider one-dimensional t, and measure the roughness of a curve $g(t)$ by its integrated squared second derivative. More general problems will be treated in Chapter 7. Here the penalized log-likelihood is:

$$\Pi = \ell(\boldsymbol{\theta}, \phi) - \tfrac{1}{2}\lambda \int g''(t)^2 dt \tag{5.22}$$

where $G(b'(\theta_i)) = \mathbf{x}_i^T \boldsymbol{\beta} + g(t_i)$, to be maximized over $\boldsymbol{\beta}$ and g, by analogy with the maximization of $S_W(\boldsymbol{\beta}, g)$ in Section 4.3.

Motivation for the penalized likelihood criterion is by now familiar. In the absence of a roughness penalty, maximization of the likelihood leads to over-fitting: $\boldsymbol{\beta}$ will not be identifiable and the curve g will fit the data exactly (apart from replicates at the same values of explanatory variables), thus becoming implausibly rough.

5.4.2 *Finding maximium penalized likelihood estimates by Fisher scoring*

We reduce the infinite-dimensional choice entailed in maximizing (5.22) to a finite one by using an argument that has appeared twice before. Let $s_1 < s_2 < \ldots < s_q$ denote the unique values among $t_1, t_2, \ldots, t_n$, arranged in increasing order. Define the incidence matrix N by $N_{ij} = 1$ if $t_i = s_j$, and 0 otherwise. Then we first maximize Π subject to

$$g(s_j) = g_j \text{ for } j = 1, 2, \ldots, q, \tag{5.23}$$

and then maximize the result over $\mathbf{g} = (g_1, g_2, \ldots, g_q)^T$. But the log-likelihood term ℓ in Π depends on g only through the values of $g(s_j)$, so the maximizing g is that *minimizing* the roughness penalty subject to (5.23), i.e. the natural cubic spline interpolating $g(s_j) = a_j$. The resulting value of the penalty is $\mathbf{g}^T K \mathbf{g}$, where $K = QR^{-1}Q^T$ as in (2.3), and Q and R are as defined in Section 2.1.2 but with $(s_1, s_2, \ldots, s_q)$ replacing $(t_1, t_2, \ldots, t_n)$.

Calculation of the MPLEs of $\boldsymbol{\beta}$ and g then reduces to the requirement to

(i) maximize $\Pi = \ell(\boldsymbol{\theta}, \phi) - \frac{1}{2}\lambda \mathbf{g}^T K \mathbf{g}$ over $\boldsymbol{\beta}$ and $\mathbf{g}$ subject to $G\{b'(\theta_i)\} = \mathbf{x}_i^T \boldsymbol{\beta} + (N\mathbf{g})_i$, and then

(ii) find the natural cubic spline g interpolating $a_j = g(s_j), j = 1, 2, \ldots, q$.

Step (i) can be carried out using Fisher scoring in a generalization of the standard parametric case that was covered in Section 5.2.

Theorem 5.2 *The Fisher scoring algorithm for maximizing the penalized log-likelihood (5.22) with respect to $\boldsymbol{\beta}$ and $\mathbf{g}$ for fixed ϕ is given by solving*

$$\begin{bmatrix} X^T W X & X^T W N \\ N^T W X & N^T W N + \alpha K \end{bmatrix} \begin{pmatrix} \boldsymbol{\beta}^{new} \\ \mathbf{g}^{new} \end{pmatrix} = \begin{pmatrix} X^T W \mathbf{z} \\ N^T W \mathbf{z} \end{pmatrix} \tag{5.24}$$

where

$$\alpha = \lambda\phi,$$

the working response vector $\mathbf{z}$ now has the form

$$z_i = (Y_i - \mu_i)G'(\mu_i) + (X\boldsymbol{\beta} + N\mathbf{g})_i, \tag{5.25}$$

and W is the diagonal matrix

$$W = \text{diag}[\{G'(\mu_i)^2 b''(\theta_i)\}^{-1}]. \tag{5.26}$$

Proof. Let $\boldsymbol{\eta}$ denote the n-vector of predictors, $\boldsymbol{\eta} = X\boldsymbol{\beta} + N\mathbf{g}$. Then

$$\frac{\partial \Pi}{\partial \boldsymbol{\beta}} = X^T \frac{\partial \ell}{\partial \boldsymbol{\eta}}$$

and

$$\frac{\partial \Pi}{\partial \mathbf{g}} = N^T \frac{\partial \ell}{\partial \boldsymbol{\eta}} - \lambda K \mathbf{g}.$$

Similarly

$$E\left(-\frac{\partial^2 \Pi}{\partial \boldsymbol{\beta} \partial \boldsymbol{\beta}^T}\right) = X^T E\left(-\frac{\partial^2 \ell}{\partial \boldsymbol{\eta} \partial \boldsymbol{\eta}^T}\right) X,$$

$$E\left(-\frac{\partial^2 \Pi}{\partial \boldsymbol{\beta} \partial \mathbf{g}^T}\right) = X^T E\left(-\frac{\partial^2 \ell}{\partial \boldsymbol{\eta} \partial \boldsymbol{\eta}^T}\right) N,$$

and

$$E\left(-\frac{\partial^2 \Pi}{\partial \mathbf{g} \partial \mathbf{g}^T}\right) = N^T E\left(-\frac{\partial^2 \ell}{\partial \boldsymbol{\eta} \partial \boldsymbol{\eta}^T}\right) N + \lambda K.$$

But the derivatives of ℓ with respect to $\boldsymbol{\eta}$ were found earlier (5.12, 5.13), so the Fisher scoring equations for simultaneously updating $\boldsymbol{\beta}$ and $\mathbf{g}$ have the block matrix form

$$\begin{pmatrix} \boldsymbol{\beta}^{new} \\ \mathbf{g}^{new} \end{pmatrix} = \begin{pmatrix} \boldsymbol{\beta} \\ \mathbf{g} \end{pmatrix} + \begin{bmatrix} X^T W X & X^T W N \\ N^T W X & N^T W N + \alpha K \end{bmatrix}^{-1} \begin{pmatrix} X^T W \mathbf{z}^* \\ N^T W \mathbf{z}^* - \alpha K \mathbf{g} \end{pmatrix}$$

and this can be rewritten in the required form (5.24). □

To summarize, each iteration in the Fisher scoring procedure for evaluating the MPLEs of β and $\mathbf{g}$ involves solving a system of $(p+q)$ linear equations. These have the same form as the partial spline estimating equations (4.6), whose solution was discussed in Sections 4.3.4 and 4.3.5. If the parametric part $\mathbf{x}_i^T\beta$ of the model is absent, and the $\{t_i\}$ are distinct and already ordered so that $q = n$ and $N = I$, then the equations reduce to

$$(W + \alpha K)\mathbf{g}^{new} = W\mathbf{z}.$$

This is the same solution as was given in Theorem 5.1 for the case of the canonical link function. In particular in the normal case $\mathbf{z} = \mathbf{Y}$, so that no iteration is needed, and we return to the cubic smoothing spline equation (3.22).

Further discussion of this method of estimation can be found in Green and Yandell (1985).

Complete algorithm

Given data $\mathbf{Y}, X, \mathbf{t}$:

Step 1 Sort $t_1, \ldots, t_n$, eliminating ties, to form $s_1, \ldots, s_q$, and the incidence matrix N.

Step 2 Initialize iteration: e.g. set $\beta = 0$,

$$\mathbf{g} = (N^T N)^{-1} N^T (G(Y_1), G(Y_2), \ldots, G(Y_n))^T.$$

Step 3 Update model: Set $\eta = X\beta + N\mathbf{g}$. Let μ, θ satisfy $G(\mu_i) = \eta_i$, $b'(\theta_i) = \mu_i$.

Step 4 Calculate the working response vector $\mathbf{z}$ and weight matrix W from η using (5.25) and (5.26).

Step 5 Use one of the methods of Sections 4.3.4 and 4.3.5 to solve (5.24) for β^{new} and $\mathbf{g}^{new}$.

Step 6 Test for convergence, and otherwise set $\beta = \beta^{new}$, $\mathbf{g} = \mathbf{g}^{new}$ and return to Step 3.

Step 7 Complete the smoothing of $\mathbf{z} - X\beta$ by obtaining all polynomial coefficients for g.

5.4.3 Cross-validation for GLMs

As each successive model has been introduced in this book, we have adapted the notion of cross-validation, by defining the appropriate score that can be used to make an automatic choice of the smoothing parameter. In this section, we attempt to do the same for semiparametric generalized linear models. This methodology is still in its infancy, and there continues to be some debate about how to set up the appropriate score. We will describe two approaches.

The first of these goes back to first principles. In Section 4.4, the cross-validation score for weighted semiparametric models was defined as

$$CV(\alpha) = \sum w_i(Y_i - \hat{Y}_i^{(-i)})^2 \tag{5.27}$$

where $\hat{Y}_i^{(-i)} = X\hat{\beta}^{(-i)} + N\hat{\mathbf{g}}^{(-i)}$. This definition encompasses all the previous cases. In a generalized linear model, to use a score based on weighted sums of squares seems unnatural, and it is therefore preferable to work in terms of the likelihood. To introduce this, note that each term of (5.27) can be interpreted as the deviance increment $d_i^{(-i)}$ of the observation Y_i from the model fit to all observations other than Y_i, under the assumption that the observations are generated from the model

$$Y_i \sim N(\mathbf{x}_i{}^T\beta + g(t_i), w_i^{-1}).$$

This is the key to definition of a score for our present, more general, models. Let

$$\begin{aligned} d_i^{(-i)} &= 2\phi[\ell_{max} - \ell(\hat{\theta}_i^{(-i)}, \phi)]_i \\ &= 2\{Y_i(\tilde{\theta}_i - \hat{\theta}_i^{(-i)}) - b(\tilde{\theta}_i) + b(\hat{\theta}_i^{(-i)})\} \end{aligned}$$

where

$$\hat{\theta}_i^{(-i)} = \theta(\hat{\beta}^{(-i)}, \hat{\mathbf{g}}^{(-i)}),$$

and

$$CV(\alpha) = \sum_{i=1}^{n} d_i^{(-i)}. \tag{5.28}$$

The cross-validation choice for α is that value minimizing $CV(\alpha)$, and the reasoning above ensures that it agrees with our previous definitions in the normally-distributed cases, except perhaps for a scale factor.

In these earlier cases, we found that the cross-validated residuals $Y_i - \hat{Y}_i^{(-i)}$ were easily related to the ordinary residuals $Y_i - \hat{Y}_i$ through the identity

$$Y_i - \hat{Y}_i^{(-i)} = \frac{Y_i - \hat{Y}_i}{1 - A_{ii}},$$

where A is the matrix defined by $\hat{\mathbf{Y}} = A\mathbf{Y}$; this was used to provide a simpler expression for $CV(\alpha)$ (for example (4.16)) that could be evaluated once a single model was fit to the data, thus not requiring the multiple model fits apparent from the definition (5.27). This relation between the residuals relies on the linear nature of the model, however, and the connection does not carry over to arbitrary generalized linear models.

By making use of the usual linearizing arguments, the approximation

$$d_i^{(-i)} \approx \frac{d_i}{(1 - A_{ii})^2}$$

can be justified, where d_i is the ordinary deviance increment. In this expression, the matrix A is specified in terms of the weight matrix W by (4.17) and (4.18); W is a function of the fitted natural parameters $\theta(\boldsymbol{\beta}, \mathbf{g})$, and it is therefore estimated by the weight matrix used in the final iteration of the Fisher scoring loop.

However, having made this approximation, it seems more logical to apply the linearizing argument to the deviance as well, to obtain the approximate cross-validation score

$$CV_{ap}(\alpha) = \sum w_i \left(\frac{z_i - (X\hat{\boldsymbol{\beta}} + N\hat{\mathbf{g}})_i}{(1 - A_{ii})^2} \right)^2, \tag{5.29}$$

where z_i are the working response variables used in the final iteration before convergence.

The scores defined in (5.28) and (5.29) were used by O'Sullivan, Yandell and Raynor (1986), and Green (1987). The latter paper sets the score up for the rather more general class of models we shall encounter in Section 6.5; for the necessary generality in that context, the concept of deleting observations from the fit needs delicate handling, which is accomplished by selectively fitting additional dummy covariates to the parametric part of the model. The cross-validated deviance increment $d_i^{(-i)}$ then arises as the *predictive discrepancy*, that is the increase in twice the log-likelihood obtained by forcing the additional coefficients back to zero.

Use of either the score (5.28) or its approximation (5.29) are quite appealing, but there seems to be no alternative in practice to recomputing the entire iterative fit for each value of α on a grid, in order to carry out the necessary minimization over α.

The second general approach to cross-validation in generalized linear models is to invert the order of (a) iteration to update parameters and (b) minimization of cross-validation score. That is, within each cycle of the Fisher scoring loop, choose α by cross-validation, as if the current linearized problem were the original subject of study. We have the expression (5.29) for the score at a fixed value of $\boldsymbol{\beta}$ and $\mathbf{g}$, which we could write as

$$CV(\alpha; \boldsymbol{\beta}, \mathbf{g}) = \sum w_i \left(\frac{z_i - (X\hat{\boldsymbol{\beta}}^{new} + N\hat{\mathbf{g}}^{new})_i}{(1 - A_{ii})^2} \right)^2,$$

not forgetting that $\mathbf{z}$ and W, which are given by (5.25) and (5.26), and hence A, depend on $\boldsymbol{\beta}$ and $\mathbf{g}$.

There is some evidence (Gu, 1992) that this second approach gives better results, but this is not as yet entirely conclusive. Apart from performance comparisons, the second approach does have the computational advantage that the analyst can make full use of existing, often highly efficient, software developed for the case of linear semiparametric models, simply embedding this within the Fisher scoring loop to handle the nonlinearity.

However the second method has the disadvantage that there is no simple characterization of the resulting cross-validatory choice of smoothing parameter, independent of the iterative method employed to linearize the problem. From this perspective, the first approach is much more attractive: Fisher scoring could be replaced by conjugate gradients, or some other more sophisticated optimization algorithm, without changing the meaning of cross-validation.

5.4.4 *Inference in semiparametric GLMs*

Unless a semiparametric model is being used in a purely exploratory fashion, it will generally be necessary to do more than simply find the maximum penalized likelihood estimates of $\boldsymbol{\beta}$ and g. In particular, quantities corresponding to the deviance, residuals, and estimates of parameter estimate variances that were discussed in Section 5.2.5 will be needed.

The appropriate equivalent number of degrees of freedom for error turns out to be ν, defined by

$$\nu = n - \text{tr}(S) - \text{tr}[\{X^T W(I - S)X\}^{-1} X^T W(I - S)^2 X]$$

calculated at convergence (Green, 1985). Here S is again

$$S = N(N^T WN + \alpha K)^{-1} N^T W.$$

This corresponds to the quantity that has been used informally in nonparametric linear models (Eubank, 1984, 1985) and generalized linear models (O'Sullivan *et al.*, 1986).

In parametric models, the 'error degrees of freedom' has a number of uses: it gives the expectation and the shape parameter of an approximating χ^2 distribution and its complement is the degree of complexity of the fitted model. It remains unclear whether these multiple interpretations extend to the semiparametric models (leaving aside their appropriateness in small samples), but our limited experience suggests that $n - \nu$ is well calibrated for measuring complexity, near to the degree of smoothing chosen by cross-validation (see Section 5.4.3). For much further discussion of these points, we refer the reader to Buja, Hastie and Tibshirani (1989), and Hastie and Tibshirani (1990, Appendix B).

Use of v in conjunction with the deviance defined in Section 5.2.5 allows the assessment of model adequacy and tests of nested models via the analysis of deviance. There remains the question of what degree of smoothing to employ in comparing the fits of two models: our preference is to base this on the more restricted model. Such tests concern the parametric part of the predictor only: tests for the significance of the nonparametric part have been developed by Cox and Koh (1986). See also Cox, Koh, Wahba and Yandell (1988).

In semiparametric versions of GLMs, there are theoretical difficulties additional to those found in Section 5.2.5. Asymptotic theory even for purely linear models has only emerged in rather special cases (Heckman, 1986; Rice, 1986; Speckman, 1988). Such results are at present restricted to the asymptotic distribution of regression estimates, and do not provide distributions for sums of squares or deviances. Recent results on approximation of the distribution of quadratic forms by χ^2 distributions may be useful in this regard (Buckley and Eagleson, 1988). In addition, data-dependent choice of the smoothing parameter leads to further mathematical difficulties.

One possible attitude to these difficulties is to ignore them, by using a semiparametric or nonparametric analysis in an exploratory fashion to suggest a particular parametric model, perhaps involving low-order polynomials, which is then fitted and subject to formal inference. This approach seems unaesthetic, and is strictly invalid as requiring 'multiple looks' at the data.

Another possible general approach would be to rely on bootstrapping to derive standard errors, etc., for semiparametric estimates. Such heavy computing takes us into a different league of computational task, and in any case it is not clear that the bootstrap can provide all the answers that we need.

We therefore, inevitably, fall back on statistics based on normal theory and quadratic approximations. Many practical questions can be handled using the deviance, including assessment of goodness-of-fit, and selection among nested models.

5.5 Application: tumour prevalence data

Dinse and Lagakos (1983) report on a logistic regression analysis of some bioassay data from a U.S. National Toxicology Program study of flame retardants. Data on male and female rats exposed to various doses of a polybrominated biphenyl mixture known as Firemaster FF-1 consist of a binary response variable, Y, indicating presence or absence of a particular nonlethal lesion, bile duct hyperplasia, at each animal's death. There are four explanatory variables: log dose, x_1, initial weight, x_2, cage position (height above the floor), x_3, and age at death, t. Our choice of this notation reflects the fact that Dinse and Lagakos commented on various possible treatments of this fourth variable. As alternatives to the use of step functions based on age intervals, they considered both a straightforward linear dependence on t, and higher order polynomials. In all cases, they fitted a conventional logistic regression model, the GLM with binomial error distribution and logit link function. They kept the data from male and female rats separate in the final analysis, having observed interactions with gender in an initial examination of the data.

Green and Yandell (1985) treated this as a semiparametric GLM regression problem, regarding x_1, x_2 and x_3 as linear variables, and t the splined variable. This approach avoids the issue of selecting a particular parametric dependence of the response on age, by letting the data speak for themselves through a nonparametric smooth curve. Decompositions of the fitted linear predictors for the male and female rats are shown separately in Figures 5.3 and 5.4, based on the Dinse and Lagakos data sets, consisting of 207 and 112 animals respectively. Note that misprints in the data as originally published were subsequently corrected (Dinse and Lagakos, 1984); it is the corrected data that are used here.

These figures are set out in an analogous way to the partial spline decompositions in Chapter 4. The top and middle plots represent $\mathbf{x}_i^T\hat{\beta}$ and $\hat{g}(t_i)$, on the same scale but with the origin displaced upwards, both plotted against t_i. In the lower part of each figure, the score residuals are displayed, again against t_i; the unusual appearance of these plots is due to the binary nature of the responses.

The semiparametric analysis gives estimates of the regression coefficient for the variable of principal interest, β_1, of -0.017 (0.103) and 0.631 (0.214) for the male and female rats respectively, with estimated standard errors in parentheses. Dinse and Lagakos obtain 0.012 (0.10) and 0.554 (0.20) respectively, using models linear in t. Thus the two analyses broadly agree, identifying a significant dependence of response on dose among the female rats, but not among the male ones.

5.6 Generalized additive models

The ideas of additive modelling touched on in Section 4.8 apply equally well in the context of the non-normal regression models of this chapter.

This leads to the *generalized additive model* defined by (5.5) and (5.6),

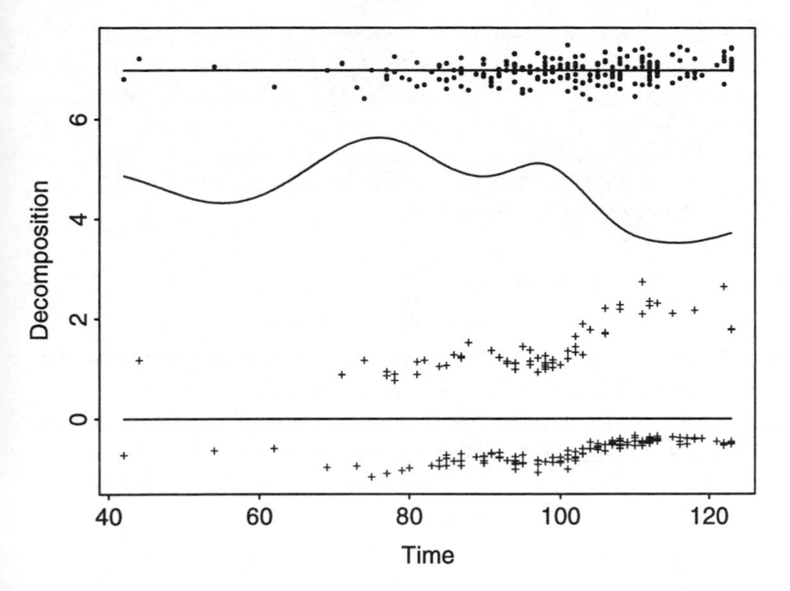

Figure 5.3. *Semiparametric logistic regression analysis for male rats*

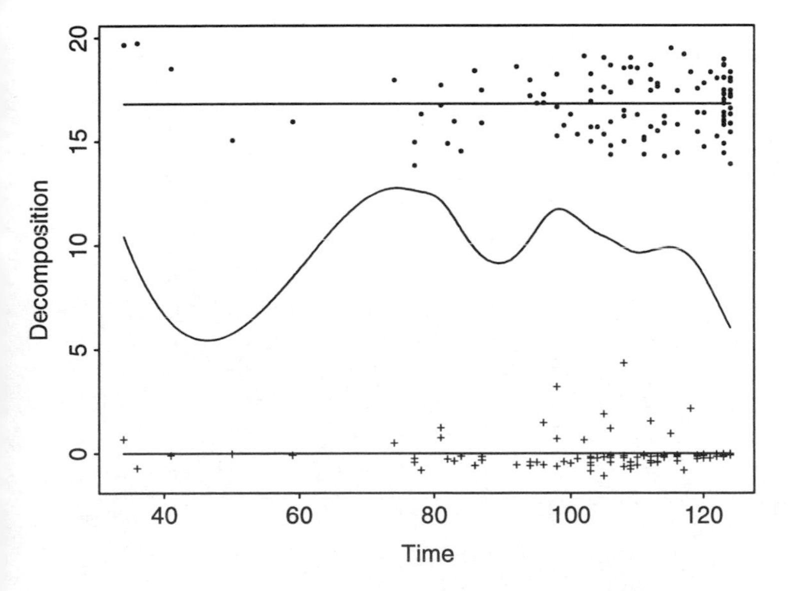

Figure 5.4. *Semiparametric logistic regression analysis for female rats*

together with

$$G(\mu_i) = \sum_{j=1}^{p} g_j(t_{ij})$$

in place of (5.7).

Derivation of maximum penalized likelihood estimators of $\{g_j\}$, to maximize the penalized log-likelihood

$$\Pi = \ell(\theta, \phi) - \tfrac{1}{2} \sum_{j=1}^{p} \lambda_j \int g_j''(t)^2 dt,$$

proceeds much as in Section 5.4.1, and the resulting estimators can be computed by backfitting within a Fisher scoring outer loop.

See Hastie and Tibshirani (1990) for a full study of these models, and Section 8.1.2 for more detail about computation.

CHAPTER 6

Extending the model

6.1 Introduction

One of the beauties of the roughness penalty approach is its conceptual versatility. In previous chapters we have considered the applications of roughness penalties in a variety of 'standard' situations, almost all of them naturally arising in linear modelling or generalized linear modelling. In this chapter we explore a number of extensions. Some of these are to more general problems, while others are to specific problems not easily or naturally dealt with by other methods.

The various problems discussed in this chapter are of course intended to be of intrinsic interest, but perhaps more importantly they provide 'templates' for the application of the roughness penalty idea to readers' own smoothing problems. In the first part of the chapter, extensions of least squares curve fitting are considered. We then move on to sections in which the generalized linear model methodology of Chapter 5 is explored further, and finally consider cases involving more general likelihoods and quasi-likelihoods.

6.2 The estimation of branching curves

In this section, we depart from standard assumptions in a rather unusual way. The application and methodology described demonstrate the usefulness of the roughness penalty approach in a very non-standard context. Fuller details are given by Silverman and Wood (1987). It should be stressed that our purpose in presenting this example is twofold. Firstly it is hoped that the methodology is of interest in its own right. Secondly the example shows the very wide potential applicability of the roughness penalty idea.

Table 6.1. *Nitrogen content (g per plant) of Sirosun 132 H plants, after Silverman and Wood (1987). The number of the branch associated with each data point is indicated*

Time of	Time of application of nitrogen				
reading	never	38 days	56 days	63 days	70 days
23 days	0.15	*	*	*	*
25 days	0.14	*	*	*	*
28 days	0.19	*	*	*	*
32 days	0.32	*	*	*	*
35 days	0.65	*	*	*	*
38 days	1.12	*	*	*	*
42 days	1.12	††	*	*	*
60 days	1.27	2.05	††	*	*
72 days	1.84	2.33	2.73	2.11	††
87 days	2.61	3.40	3.74	3.41	3.15
107 days	2.34	2.94	3.73	3.33	3.09
Branch	0	1	2	3	4

6.2.1 *An experiment on sunflowers*

Steer and Hocking (1985) carried out an experiment to test the effect of applying nitrogen to sunflowers at different stages of growth. In one of the treatments, the control, no nitrogen was applied; in the other four a nitrogen compound was applied at a given time after sowing, 38, 56, 63 and 70 days respectively. At various times the nitrogen content of plants taken from the plots was measured destructively. Up to the time of treatment, there is no difference between the treatment and the control. It is of interest to present the experiment in a way that is easily understood, and to explore the relation between the time of application of nitrogen and the overall development of the plants.

The data collected are given in Table 6.1. A starred entry in the table corresponds to a reading that can be assumed to be equal to the control, because it is before the treatment time for its column; an entry marked †† corresponds to a time at which no data were collected on the relevant treatment.

6.2.2 *The estimation method*

An elegant way of analysing these data is to model them by a branching curve, in which the estimated responses for the various treatments only diverge from the control at the point where the treatment is applied. Such a presentation immediately makes the structure of the experiment clear, and also indicates the relative efficacy of the treatment applied at various times.

The full mathematical details of the method are given in Silverman and Wood (1987). To summarize, we let $\mathcal{G}$ be a space of branching curves of a given structure with branches at the specified points. To be specific to our particular example, let $\tau_1, \tau_2, \tau_3, \tau_4$ be the four treatment times 38, 56, 63 and 70 respectively. A member g of $\mathcal{G}$ would consist of five smooth curves, a 'control curve' g_0 defined on $[23, 107]$, and, for $k = 1, 2, 3, 4$, 'treatment curves' g_k defined on $[\tau_k, 107]$ and satisfying the continuity conditions $g_k(\tau_k) = g_0(\tau_k)$. We shall refer to the individual g_i as *branches* of g, and to the points τ_k as *branch points*. A branch point will be said to lie on a particular branch if it is involved in any of the continuity conditions referring to that branch. The roughness of the branching curve, denoted by $\int g''^2$, is defined to be $\sum_i \int g_i''^2$, where the integrals are taken over the range of definition of the various functions.

It is also possible, if one wishes, to impose further continuity conditions: for example the condition $g_k'(\tau_k) = g_0'(\tau_k)$ for any particular k would yield a branching system where g_k joins g_0 smoothly at τ_k. We shall not consider this possibility in any detail, but refer the reader to Silverman and Wood (1987) for further discussion.

The observed data can be considered as triples $(i(j), t_j, Y_j)$, where t_j is the time at which the reading is taken, Y_j is the value of the reading, and $i(j)$ is the branch on which the reading lies, given by the label at the bottom of the relevant column in Table 6.1. An immediate analogue of the usual spline-smoothing nonparametric regression is then to estimate the branching curve underlying the data by the minimum $\hat{g}$ over $\mathcal{G}$ of

$$S(g) = \sum_j \{Y_j - g_{i(j)}(t_j)\} + \alpha \int g''^2. \tag{6.1}$$

Now let $\mathcal{G}_0$ be the space of *branching splines*, members of $\mathcal{G}$ for which every branch is a cubic spline with knots at its data points and branch points. It can be shown that the minimizer of $S(g)$ over $\mathcal{G}$ is necessarily a branching spline. To see this, suppose that g is any branching curve, and let $\bar{g}$ be the branching spline made up by replacing each branch g_i by the interpolating natural cubic spline to the values of g_i at its data points and branch points. Just as in Section 2.3.1, the residual sum of squares will not be affected by the replacement, but the roughness will in general be reduced.

The minimizing branching spline is found, as in the case of ordinary smoothing splines, by solving a suitable system of linear equations. Any member of $\mathcal{G}_0$ can be parametrized by specifying the value and second derivative of each branch at each of its branch points and data points. These parameters are subject to various linear constraints, as follows. At each interior knot of each of the branches, a linear condition of the form (2.27) will hold, in order to ensure continuity of the first derivative; the continuity conditions between branches will translate into identification of the values at the branch points; since each branch is a natural cubic spline the second derivative will be zero at the end knot of each branch (even if the end is an interior knot of some other branch).

By using the obvious generalization of equation (2.5), the roughness $\int g''^2$ can be expressed in terms of the parametrization, as can the residual sum of squares. Minimizing this quadratic form subject to the various linear constraints yields the branching spline that minimizes $S(g)$. It is possible to arrange the calculation in such a way that ultimately one solves a system of linear equations none of which has more than seven non-zero coefficients. The use of sparse matrix methods then yields a fast algorithm for the solution; see Silverman and Wood (1987) for details.

6.2.3 Some results

Of course, the amount by which the data are smoothed out by the procedure is determined by the value of α. If $\alpha = \infty$ then we obtain a fit (Figure 6.1) in which all the branches are straight lines, while if $\alpha = 0$ we get a smooth branching system (Figure 6.2) that interpolates the data. It should be noted that for this data structure even the interpolation problem is not at all straightforward. A clearer picture is in any case given by choosing an intermediate value of α and hence getting a smooth branching curve estimate, as in Figure 6.3. Just as in ordinary spline smoothing, the smoothing parameter could be chosen by a cross-validation approach, but in this case an external estimate of the error variance is available. The smoothing parameter value used, $\alpha = 870$, was chosen by matching this estimate to an internal estimate, in a way described by Silverman and Wood (1987). It can be seen from the figure that applying nitrogen very early in the growth cycle provides an effect that dies away later, presumably because the plants are insufficiently developed to derive full benefit. If the aim is to maximize the overall nitrogen content, then the best time to apply the treatment is at about 60 days after sowing. Perhaps an even more important feature of Figure 6.3 is that it gives a clear and easily comprehensible presentation of the overall pattern of the experiment.

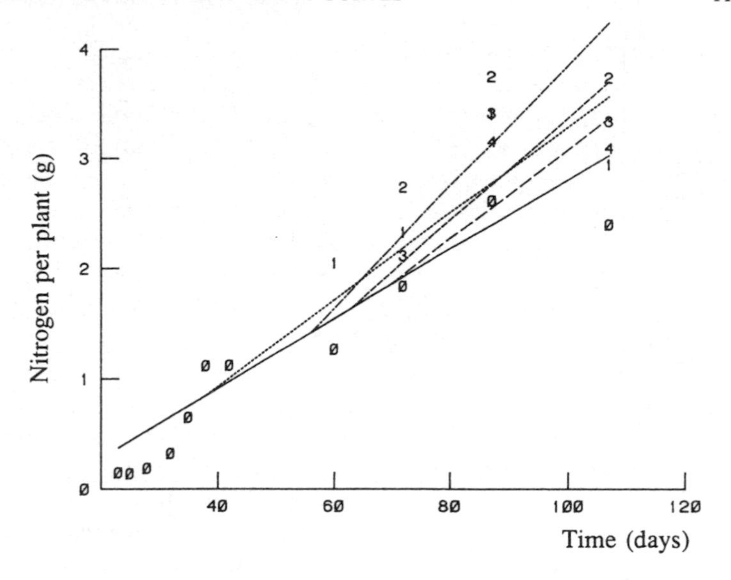

Figure 6.1. *Branching piecewise linear fit to sunflower data. Reproduced from Silverman and Wood (1987) with the permission of the American Statistical Association.*

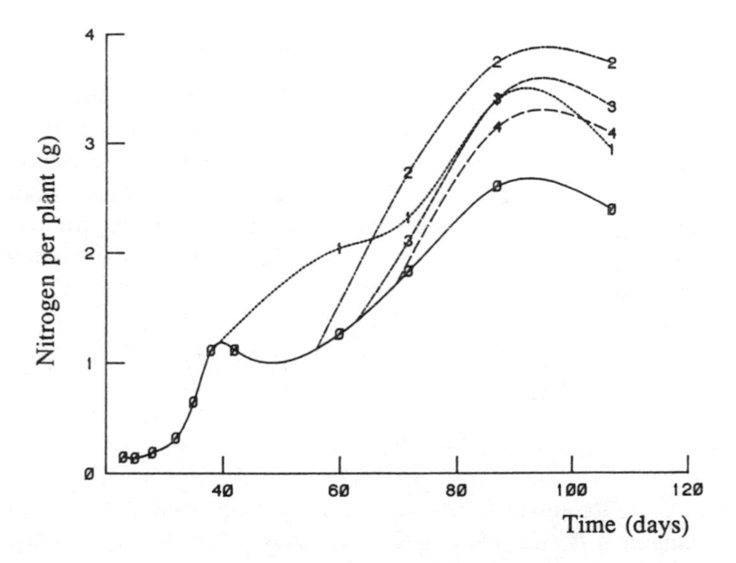

Figure 6.2. *Interpolating branching spline fit to sunflower data. Reproduced from Silverman and Wood (1987) with the permission of the American Statistical Association.*

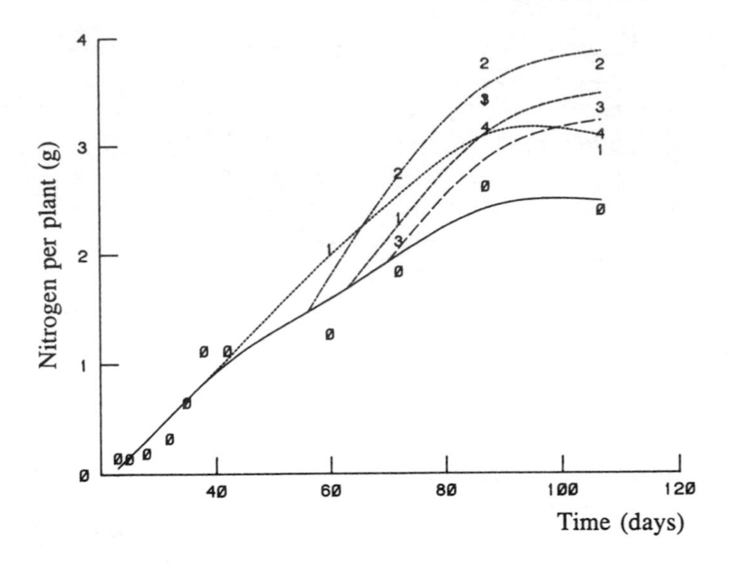

Figure 6.3. *Smoothed branching spline fit to sunflower data. Reproduced from Silverman and Wood (1987) with the permission of the American Statistical Association.*

6.3 Correlated responses and non-diagonal weights

In certain applications, especially where observations are taken sequentially in time or space, the successive errors in our model (4.3) may be positively correlated. For this and other reasons, we may occasionally wish to allow a more general pattern of weights than that appearing in the penalized sum of squares (4.4). In matrix terms, the weights matrix W will be symmetric and positive-definite but no longer diagonal, and the penalized sum of squares becomes

$$S_W(\boldsymbol{\beta}, g) = \sum_{i=1}^{n}\sum_{j=1}^{n} w_{ij}\{Y_i - \mathbf{x}_i^T\boldsymbol{\beta} - g(t_i)\}^T\{Y_j - \mathbf{x}_j^T\boldsymbol{\beta} - g(t_j)\} + \alpha \int g''(t)^2 dt. \tag{6.2}$$

Some caution should be shown before adopting this as a starting point for estimating $\boldsymbol{\beta}$ and g. It is quite well known that in fully parametric generalized least squares, estimates of $\boldsymbol{\beta}$ based on incorrect weights can be less efficient than those resulting from an unweighted analysis. This can apply even if the weights are estimated from the data. In addition, when we include a smooth curve $g(t)$ in the model there is an obvious danger that an error structure exhibiting positive autocorrelation may be partly confounded with a smoothly varying term in the systematic part of the model.

Notwithstanding these caveats there will be an occasional need for procedures that estimate β and g by minimizing (6.2), and so these are briefly discussed here. All of the relevant linear algebra continues to hold when the matrix W is not diagonal, and both the algorithms described in Sections 4.3.4 and 4.3.5 are still applicable. Both will still be computationally efficient if $S = N(N^T WN+\alpha K)^{-1}N^T W$ can be applied to a vector, for example to form $S\mathbf{Y}$ from $\mathbf{Y}$, in $O(n)$ time. The presence of the incidence matrix N adds only notational complications; we will omit it here, so that $S = (W+\alpha K)^{-1}W = (W+\alpha QR^{-1}Q^T)^{-1}W$. Now Q and R are banded, but $QR^{-1}Q^T$ is not: the trick embodied in the Reinsch algorithm is essentially equivalent to rewriting S as $I-\alpha W^{-1}Q(R+\alpha Q^T W^{-1}Q)^{-1}Q^T$ (see equations (3.23) and (3.24)). If W is diagonal, then $Q^T W^{-1}Q$ *is* banded, and the algorithm works in $O(n)$ time.

Departures from the assumption of uncorrelated errors in regression are often handled by using the correlation structures of simple stationary time series models. If the errors follow a *moving average* process of low order, then W^{-1} is not diagonal, but is still banded. Thus $Q^T W^{-1}Q$ is banded, although it now has more non-zero bands than before. The Reinsch algorithm is again applicable, with only modest changes.

On the other hand, if an *autoregressive* structure for the errors is assumed, then it is W, not its inverse, that is banded. A different approach is then needed, going a little further back towards first principles. Equations (3.23) and (3.24) can be re-expressed in the symmetric block matrix form:

$$\begin{bmatrix} \alpha^{-1}W & Q \\ Q^T & -R \end{bmatrix} \begin{pmatrix} \mathbf{g} \\ \gamma \end{pmatrix} = \begin{pmatrix} \alpha^{-1}W\mathbf{Y} \\ 0 \end{pmatrix}.$$

Now each of the blocks in these equations is banded but not diagonal. If we were to permute the rows and columns of the $(2n-2)\times(2n-2)$ matrix by taking them in the order: $1, 2, (n+1), 3, (n+2), 4, \ldots, (2n-2), n$ then it is quite easy to see that the result would be banded. The bandwidth is $\max(2b-1, 7)$ where b is the bandwidth of W. The procedures using the Cholesky decomposition that were described in detail in Section 2.6 can then be adapted to this new problem.

6.4 Nonparametric link functions

We now leave linear models, and consider a different 'semiparametric' version of GLMs from that of Section 5.4. This is obtained by reverting to the ordinary GLM of Section 5.2, but supposing that in addition to the regression parameter $\boldsymbol{\beta}$, the link function G is also unknown. Of course, in truth, this is almost always the case, except when a very definite model for the mechanism generating the data is available, but the usual practice is to fix G nevertheless, and to concentrate on the estimation of $\boldsymbol{\beta}$.

If estimating G is of interest, one possible approach is the parametric link function method of Scallan, Gilchrist and Green (1984). However it is more in the spirit of this book to assume that G is completely unknown, but for smoothness assumptions, and to estimate it nonparametrically simultaneously with $\boldsymbol{\beta}$; in this section we discuss such an approach.

It turns out to be more straightforward to work not directly with the link function defined in (5.7), but with a function related to its inverse. This will avoid the need to impose awkward constraints to guarantee the existence of certain inverse functions to make our model well-defined. We will therefore assume that our responses are drawn independently from the density or probability function (5.5), where the natural parameter θ_i for the ith observation is related to the linear predictor by

$$\theta_i = g(\mathbf{x}_i^T \boldsymbol{\beta}). \tag{6.3}$$

Both g and $\boldsymbol{\beta}$ are unknown, and in the spirit of our general approach, we could attempt to estimate them both by maximizing the penalized likelihood

$$\ell(\boldsymbol{\theta}, \phi) - \tfrac{1}{2}\lambda \int g''(\eta)^2 d\eta \tag{6.4}$$

along the lines of (5.22). Notice, however, that the integration is now in the space of the linear predictor η, not an explanatory variable t. An appealing property of this criterion combined with this parametrization of the model is that perfectly smooth g (those for which $\int g''^2 = 0$) correspond to choice of the canonical link function.

Inspection of (6.3) reveals a certain lack of identifiability. For example if $\boldsymbol{\beta}$ is replaced by $\tilde{\boldsymbol{\beta}}$ for some scalar c, and g is replaced by $\tilde{g}$, where $\tilde{g}(\eta_i) = g(c^{-1}\eta_i)$, then $\boldsymbol{\theta}$ is unchanged. Similarly, if the linear predictor includes a constant term, there is a translation equivariance as well. A full analysis of such equivariances is quite complicated, and will not be attempted here. Instead, we will only consider models in which the linear predictor *does* include a constant term, and constrain the parameter vector $\boldsymbol{\beta}$ so that the linear predictor is standardized to have mean 0 and variance 1, for the particular set of explanatory variables $\mathbf{x}_i$ to hand. It is evident that this can always be imposed by an explicit scaling of $\boldsymbol{\beta}$, and a shift of the constant term. The rationale for this pair of constraints is that these are precisely the degrees of freedom absorbed by the linear regressions that form the perfectly smooth functions g.

The estimating equations that arise from maximizing (6.4) subject to the constraints

$$\sum \eta_i = 0 \quad \text{and} \quad \sum \eta_i^2 = (n-1) \tag{6.5}$$

have awkward Lagrange multiplier terms. This route will not be pursued here. Instead we will simply alternate between maximizing (6.4) over $\boldsymbol{\beta}$ for fixed g, and over g for fixed $\boldsymbol{\beta}$. After each update of $\boldsymbol{\beta}$, $\boldsymbol{\eta}$ and g will be rescaled as described above to impose (6.5), while leaving $\boldsymbol{\theta}$ unchanged.

For fixed g, the model is an ordinary parametric GLM, so maximization of (6.4) is an application of the methods of Section 5.2. An updated estimate of $\boldsymbol{\beta}$ is given, as in (5.9) by

$$\boldsymbol{\beta}^{new} = (X^T W^{(1)} X)^{-1} X^T W^{(1)} \mathbf{z}^{(1)},$$

where in the present notation,

$$z_i^{(1)} = \frac{Y_i - \mu_i}{b''(\theta_i) g'(\eta_i)} + \mathbf{x}_i^T \boldsymbol{\beta},$$

and $W^{(1)}$ is the diagonal matrix with

$$W_{ii}^{(1)} = b''(\theta_i)\{g'(\eta_i)\}^2.$$

Conversely for $\boldsymbol{\beta}$ and hence $\boldsymbol{\eta}$ fixed, our model is a purely nonparametric GLM, using the canonical link function, with η_i as the values of a one-dimensional working explanatory variable, so the methods of Section 5.3 are relevant. The update is given by (5.20), which in the present context becomes

$$\mathbf{g}^{new} = (W^{(2)} + \alpha K)^{-1} W^{(2)} \mathbf{z}^{(2)},$$

with

$$z_i^{(2)} = \theta_i + \frac{Y_i - b'(\theta_i)}{b''(\theta_i)}$$

and

$$W_{ii}^{(2)} = b''(\theta_i).$$

Thus no really new ideas are needed to handle the nonparametric link function model: this is another example of the flexible, modular way in which computational tools arising from roughness penalty methods can be combined to produce new methodology.

Before giving an example, we close this section with a word of warning. As with all of the iterative algorithms for nonlinear models in this book, there is no guarantee of convergence from all initial estimates. In most cases, the algorithms are well-behaved for most data sets, but it does seem that the algorithm above is prone to limit-cycle behaviour in some cases, typically when the smoothing parameter is such as to allow more than about 5 degrees of freedom for the nonparametric curve. In our

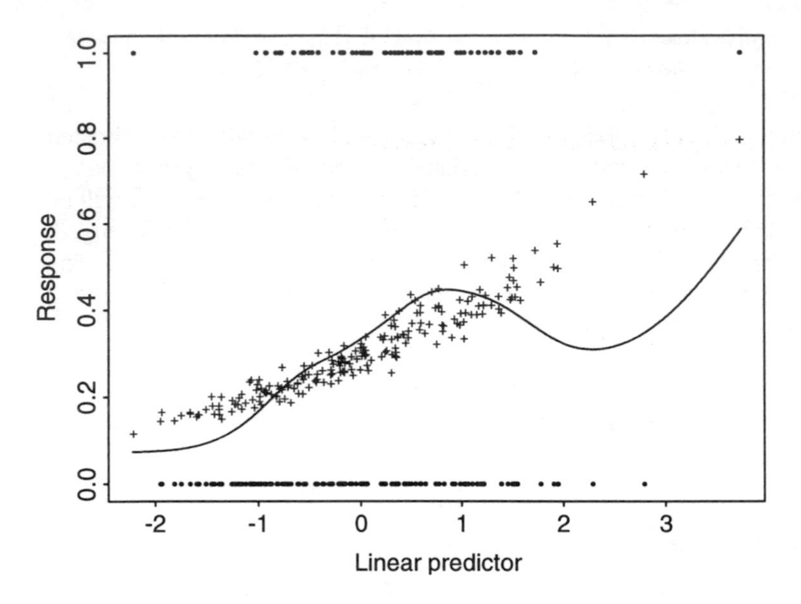

Figure 6.4. *Nonparametric link function analysis, allowing 5 d.f., for the male rats in the Dinse and Lagakos study. Smooth curve shows nonparametric fitted probabilities, + signs those obtained from a standard logistic regression, and points the raw data.*

experience, this problem can be circumvented by adjusting the step size in both of the updating steps, that is, by updating to a convex combination of the previous value and that suggested by the algorithm as described. Since both β and g are updated twice in each cycle, because of the rescaling, it is of course the previous value *at that point in the cycle* that should be used here.

6.4.1 Application to the tumour prevalence data

As a demonstration of the methodology introduced above, we show in Figure 6.4 the results of applying it to the data on male rats from the tumour prevalence study described in Section 5.5. The number of degrees of freedom for the smooth link function was fixed at 5, and the linear predictor included terms corresponding to all four of the explanatory variables (x_1, x_2, x_3 and t), all featuring linearly. Figure 6.4 displays the observed responses Y_i, plotted against the fitted values of the linear predictor $\hat{\eta}_i$, and the smooth fitted probability curve $\exp\{\hat{g}(\eta)\}/[1+\exp\{\hat{g}(\eta)\}]$.

In order to provide a comparison with a standard analysis, the fitted probabilities from a fully parametric logistic regression on x_1, x_2, x_3 and t are also displayed, plotted against the *non*parametric fitted $\hat{\eta}_i$.

6.5 Composite likelihood function regression models

We now turn to extensions of the generalized linear model structure discussed in Chapter 5. In this section, we relax further some of the specific assumptions used by Nelder and Wedderburn (1972) to define generalized linear models. In particular, we will no longer insist on the exponential family of distributions, on the independence of the responses, or on the linearity of the regression function. What will be kept is the pivotal role of the *predictors* $\{\eta_i\}$ in decomposing the regression model into systematic and random components. Among examples that can then be covered that lie outside the scope of generalized linear models are

- multiple linear regression with an arbitrary error distribution,
- Gaussian nonlinear regression,
- models for censored or grouped data,
- generalizations of probit analysis to allow synergistic or antagonistic interaction between drugs, and
- multinomial models for regression analysis of ordered categorical data.

This framework for regression has been further developed by Green (1984) and Green (1987), particularly from an algorithmic perspective.

We suppose that we are given a regression function that determines the predictors $\boldsymbol{\eta} = (\eta_1, \eta_2, \ldots, \eta_n)^T$ in terms of an unknown p-vector of parameters $\boldsymbol{\beta}$, an unknown function g of a single variable t, and observed explanatory variables. Suppose that $\boldsymbol{\eta} = \boldsymbol{\eta}(\boldsymbol{\beta}, g)$ depends on g only through the values of $g(t_1), g(t_2), \ldots, g(t_m)$, where $t_1, t_2, \ldots, t_m$ are known. Often m will be the same as n, although this is not necessary, and $t_1, t_2, \ldots, t_n$ may then be observed values of an explanatory variable for each of n cases.

This regression function provides the systematic component of the model, and the random component is provided by the log-likelihood

$$\ell = \ell(\mathbf{Y}; \boldsymbol{\eta}, \phi) = \ell(\boldsymbol{\eta}, \phi), \text{ say,}$$

of the data $\mathbf{Y}$ in terms of the vector of predictors $\boldsymbol{\eta}$. As in Section 5.4, we are interested in estimating $\boldsymbol{\beta}$ and g, in order to quantify the dependence of the responses $\mathbf{Y}$ on the explanatory variables $\mathbf{x}$ and t. We do so by maximizing the penalized log-likelihood

$$\Pi = \ell\{\boldsymbol{\eta}(\boldsymbol{\beta}, g), \phi\} - \tfrac{1}{2}\lambda \int g''(t)^2 dt. \tag{6.6}$$

The extra generality actually causes very few new difficulties. Suppose for notational simplicity that there is no nuisance parameter ϕ parametrizing the random component of the model or, which amounts to the same thing, that $\phi = 1$. The restrictions placed on the dependence of η on g ensure that once again the estimate of g is guaranteed to be a natural cubic spline and we have only to solve a finite-dimensional optimization problem. Once again we attempt an iterative solution by Fisher scoring, and evaluate the requisite derivatives of Π by the chain rule. Let $\mathbf{g} = (g(s_1), ..., g(s_q))^T$ where $s_1, s_2, ..., s_q$ denote the ordered distinct values among $t_1, t_2, ..., t_m$. Then define

$$X = \frac{\partial \eta}{\partial \beta}, \quad N = \frac{\partial \eta}{\partial \mathbf{g}} \tag{6.7}$$

and

$$W = E\left(-\frac{\partial^2 \ell}{\partial \eta \partial \eta^T}\right); \tag{6.8}$$

let

$$\mathbf{z} = W^{-1}\frac{\partial \ell}{\partial \eta} + X\beta + N\mathbf{g}. \tag{6.9}$$

With these changes of definition, all of the arguments of Section 5.4 still apply and we again obtain the updating equations (5.24).

We must distinguish various cases when it comes to the practical implementation of these updating equations. If W is diagonal and N still an incidence matrix then the algorithm of Section 5.4.2 applies with only trivial changes:

Step 2′ Initialize β and $\mathbf{g}$ appropriately (there is no longer a link function to make this explicit).

Step 3′ Calculate $\eta = \eta(\beta, \mathbf{g})$ using the appropriate regression function. Compute X and N given by (6.7).

Step 4′ Calculate W and $\mathbf{z}$ from (6.8) and (6.9).

If either W or N has sufficiently more complicated structure that $N^T WN$ is not diagonal, then the Reinsch algorithm is no longer applicable. One of the approaches outlined in Sections 6.3 or 8.2 can then be used instead.

6.6 A varying coefficient model with censoring

To give some of the flavour of the models of the preceding section, we consider here a rather contrived example in which appear the ingredients of semiparametric modelling, censoring, an error distribution outside the exponential family, and the varying coefficient models introduced and studied by Hastie and Tibshirani (1993). This combination by no means exhausts the scope of composite likelihood regression models!

Suppose that potential responses $\{\tilde{Y}_i, i = 1, 2, \ldots, n\}$ are independently drawn from a location family

$$P\{\tilde{Y}_i \leq y\} = \Psi(y - \eta_i) \tag{6.10}$$

where Ψ is a known distribution function with density ψ, and the predictor η_i is given by

$$\eta_i = \mathbf{x}_i^T \boldsymbol{\beta} + s_i g(t_i). \tag{6.11}$$

Here, $\mathbf{x}_i$, s_i and t_i are observed explanatory variables (a p-vector and two scalars, respectively) and the parameter $\boldsymbol{\beta}$ and function g are, as usual, to be estimated. Note that the nonparametric function g appears here as the *slope* of the regression on the variable s; its argument t is called an *effect modifying variable* by Hastie and Tibshirani.

The potential responses are subject to censoring, and we actually observe only

$$Y_i = \min(\tilde{Y}_i, c_i)$$

and

$$\delta_i = \begin{cases} 1 & \text{if} \quad Y_i < c_i \\ 0 & \text{if} \quad Y_i = c_i, \end{cases}$$

where c_i is the non-random censoring time for the ith observation.

The log-likelihood thus takes the form

$$\ell(\boldsymbol{\eta}) = \sum_{i=1}^{n} [\delta_i \log \psi(Y_i - \eta_i) + (1 - \delta_i) \log\{1 - \Psi(Y_i - \eta_i)\}] \tag{6.12}$$

and the maximum penalized likelihood estimates of $\boldsymbol{\beta}$ and g are those maximizing

$$\ell(\boldsymbol{\eta}) - \tfrac{1}{2}\lambda \int g''(t)^2 dt$$

as usual. This is an example of (6.6). Construction of an algorithm to carry out the estimation proceeds as in Section 6.5, with one exception. Since it is not usually reasonable to assume that censoring times $\{c_i\}$ are known for *uncensored* observations, the expectation in (6.8) cannot be computed. We therefore take instead

$$W = -\frac{\partial^2 \ell}{\partial \boldsymbol{\eta} \partial \boldsymbol{\eta}^T}$$

that is, use observed rather than expected information in the scoring step. This substitution does not usually materially affect the convergence of the scoring algorithm; for further discussion of this point, see Jørgensen

(1983). The matrix W is diagonal, consisting of weights that arise by straightforward but tedious differentiation of (6.12).

The derivative matrices X and N are easily obtained for this model. The rows of X are just the vectors of explanatory variables $\mathbf{x}_i^T$, while it is clear from (6.11) that

$$N = UN^*,$$

where N^* is the incidence matrix describing the ordering and ties among $\{t_1, t_2, \ldots, t_n\}$, as in Section 4.3.1, and

$$U = \text{diag}(s_1, s_2, \ldots, s_n).$$

Fitting the model can proceed by the backfitting method of Section 4.3.4. Following (5.24), we alternate between use of the equations

$$\boldsymbol{\beta} = (X^T W X)^{-1} X^T W(\mathbf{z} - N\mathbf{g}),$$

and

$$\mathbf{g} = S(\mathbf{z} - X\boldsymbol{\beta}),$$

where

$$S = N(N^T W N + \alpha K)^{-1} N^T W.$$

Since $N = UN^*$, we can write

$$S = U\{N^*(N^{*T} UWUN^* + \alpha K)^{-1} N^{*T} UWU\} U^{-1}$$

so that the smoothing operation consists of weighted cubic spline smoothing allowing for ties and using weight matrix UWU, preceded and followed, respectively, by division and multiplication by U.

Various adaptations of this model could be contemplated. For example, we might wish to replace the additive predictor (6.11) by a function nonlinear in $\boldsymbol{\beta}$ and g, such as

$$\eta_i = \beta_1 x_i^{\beta_2} \beta_3^{(x_i^{\beta_4})} [1 - \exp\{-s_i g(t_i)\}]^{\beta_5} x_i^{\beta_6},$$

(a modification of a model that has been used to describe population growth). This would not change the algorithm, but merely mean that the matrices X and U would have more complicated forms, and values that changed with each iteration. Similarly, replacing the location family (6.10) by a more general dependence

$$P\{\tilde{Y}_i \le y\} = \Psi(y; \eta_i)$$

would simply mean a different calculation for W and $\mathbf{z}$.

In all of these variations, the fixed point justification of the backfitting algorithm for maximum penalized likelihood estimation of $\boldsymbol{\beta}$ and g is retained. More care may be needed, however, in the more exotically nonlinear models, and it may also be necessary for convergence that the step size is adjusted, for example by taking a fixed convex combination of the old and proposed new values of each parameter at each update.

6.7 Nonparametric quantile regression

In medical practice, a useful device for screening patients is a table or chart recording the distribution of a biometrical measurement, such as height, weight or middle-upper-arm-circumference, for various values of an appropriate covariate, often age. A physician can then refer a measurement on a current patient to this chart, in order to assess, for example, whether a child is unusually short or tall for his age. Naturally, the chart has to be constructed from data from a relevant population.

In most cases of interest, the covariate concerned takes continuous values, and the distribution of the measurement can be assumed to vary smoothly with the covariate. Reference charts should therefore be constructed so as to respect this continuity, and without making use of arbitrary discretization or grouping. Further, although it is sometimes argued on general grounds that biometrical measurements are approximately normally distributed, it seems more appropriate not to make strong distributional assumptions, or to assume a particular parametric form for the dependence on the covariate.

6.7.1 The LMS method

A variety of methods is available for constructing reference curves for populations, not all of them meeting all the criteria set out above. A method that does meet them was proposed by Cole (1988). In what he terms the LMS method, it is assumed that for each value of the covariate t, the measurement Y, *after a Box–Cox power transformation*, is normally distributed. It is supposed that the expectation, variance, and the power of the transformation, all vary smoothly with t. Thus there is flexibility over the way that the location, spread and shape of the distribution of Y given t depend on t; it is natural, and in the spirit of this monograph, to allow this dependence to be nonparametric. Because of the flexibility afforded by these three varying characteristics, the normal assumption is not a particularly strong one.

The methodology described here was originally developed with the application to reference growth curve data very much in mind, but of course it may have applications in other areas where it is required to produce, as a function of a covariate, an estimate not just of the expected response, but of the whole distribution.

To define the LMS method, suppose that $\lambda(t)$, $\mu(t)$ and $\sigma(t)$ are functions of t, with $\mu(t)$ and $\sigma(t)$ supposed strictly positive. It is the initial letters of λ, μ and σ that provide the acronym LMS. If a measurement Y is made at t, denote by Z the transformed variate

$$Z = \begin{cases} \sigma(t)^{-1}\lambda(t)^{-1}[\{Y/\mu(t)\}^{\lambda(t)} - 1] & \text{if} \quad \lambda(t) \neq 0 \\ \sigma(t)^{-1}\log\{Y/\mu(t)\} & \text{if} \quad \lambda(t) = 0 \end{cases}$$

Then the conditional distribution of Z given t is assumed to be standard normal. If Φ denotes the normal distribution function, then the $100\alpha\%$ quantile of Y at t will be

$$\mu(t)\{1 + \lambda(t)\sigma(t)\Phi^{-1}(\alpha)\}^{1/\lambda(t)} \qquad \text{for} \quad \lambda(t) \neq 0$$

or

$$\mu(t)\exp\{\sigma(t)\Phi^{-1}(\alpha)\} \qquad \text{for} \quad \lambda(t) = 0.$$

Thus if the λ, μ and σ curves are smooth, so are all quantile curves. Cole's proposal is to use a plot of selected quantile curves as a summary of reference growth data.

6.7.2 *Estimating the curves*

There are various ways in which the λ, μ and σ curves could be estimated from data. Here, we describe the approach suggested by Cole and Green (1992), which is appropriate in the absence of assumptions about the forms of the curves other than that they are smooth. As described here, the method applies to data consisting of independent observations (t_i, y_i), $i = 1, 2, \ldots, n$, such as would arise in cross-sectional studies, and a modification to the likelihood would be needed to deal with the (more common) case of longitudinal data.

The penalized log-likelihood is

$$\begin{aligned} \Pi &= \sum_{i=1}^{n} \left(\lambda(t_i) \log \frac{Y_i}{\mu(t_i)} - \log \sigma(t_i) - \tfrac{1}{2} Z_i^2 \right) \\ &\quad - \tfrac{1}{2}\alpha_\lambda \int \lambda''(t)^2 dt - \tfrac{1}{2}\alpha_\mu \int \mu''(t)^2 dt - \tfrac{1}{2}\alpha_\sigma \int \sigma''(t)^2 dt, \end{aligned}$$

where α_λ, α_μ and α_σ are smoothing parameters, and Z_i denotes Y_i transformed as above. A minor difficulty now is that there is a singularity in the likelihood as $\sigma(t) \to 0$ for all t. This gives an infinite value for Π. However, there is in our experience with the model always a local maximum of Π which is a stable fixed point of the algorithm about to be described, and to which the algorithm converges from 'sensible' initial estimates.

To locate a local maximum of Π, we now follow the familiar paradigm of using Fisher scoring. Let $\boldsymbol{\lambda}$, $\boldsymbol{\mu}$ and $\boldsymbol{\sigma}$ denote the vectors of values of the corresponding curves at the distinct ordered values of $\{t_i\}$. Given $\boldsymbol{\lambda}$, $\boldsymbol{\mu}$ and $\boldsymbol{\sigma}$ at the current iteration, updated estimates are given by the solutions to the equations

$$\begin{bmatrix} W_\lambda + \alpha_\lambda K & W_{\lambda\mu} & W_{\lambda\sigma} \\ W_{\mu\lambda} & W_\mu + \alpha_\mu K & W_{\mu\sigma} \\ W_{\sigma\lambda} & W_{\sigma\mu} & W_\sigma + \alpha_\sigma K \end{bmatrix} \begin{pmatrix} \boldsymbol{\lambda}^{new} - \boldsymbol{\lambda} \\ \boldsymbol{\mu}^{new} - \boldsymbol{\mu} \\ \boldsymbol{\sigma}^{new} - \boldsymbol{\sigma} \end{pmatrix}$$

$$= \begin{pmatrix} \mathbf{u}_\lambda - \alpha_\lambda K \boldsymbol{\lambda} \\ \mathbf{u}_\mu - \alpha_\mu K \boldsymbol{\mu} \\ \mathbf{u}_\sigma - \alpha_\sigma K \boldsymbol{\sigma} \end{pmatrix}.$$

The W's and $\mathbf{u}$'s are first and expected second derivatives of the log-likelihood with respect to the variables specified in the subscripts, and are sums of terms over observations at each distinct value of t. (In the case of W_λ, the expectation of the second derivative is replaced by that of the first few terms of a Taylor approximation, as derivatives of the log-likelihood with respect to $\boldsymbol{\lambda}$ do not have finite expectations.) This system of equations, while awkward to handle directly, is very amenable to backfitting: each of the three block rows of the system can be rearranged in the form of an updating equation for $\boldsymbol{\lambda}$, $\boldsymbol{\mu}$ and $\boldsymbol{\sigma}$ respectively, which is an application of a weighted cubic spline smoother. Cole and Green find that this algorithm converges in typically four to eight iterations of the outer scoring loop. The iteration can be initialized by setting $\boldsymbol{\lambda}$ identically to 1, obtaining $\boldsymbol{\mu}$ by a simple smooth of $\mathbf{Y}$, and $\boldsymbol{\sigma}$ by smoothing the vector of values $\{Y_i/\mu(t_i) - 1\}^2$.

Further details of the methodology can be found in Cole and Green (1992), which includes a discussion of the relationship between the smoothing parameters and the corresponding equivalent degrees of freedom for each curve, and illustrations of the LMS method applied to data on triceps skinfold in Gambian females, and on body weight in U.S. girls.

Here, we provide some illustrations from the Gambian study, also including figures on males not included in Cole and Green (1992). Figure 6.5 shows a scatter plot of triceps skinfold on age for Gambian females aged from about 3 to 26 years. The percentile curves shown are estimated by the method we have described, using smoothing parameters that allow 6 equivalent degrees of freedom for each curve. Note the 'notch' in the dependence at around 9 years. Figure 6.6 displays the fitted λ curve for these data, and superimposed the fitted σ curve. The distribution of triceps skinfold is apparently positively skewed for ages less than 20 years, with the skew decreasing abruptly between 20 and 25, although

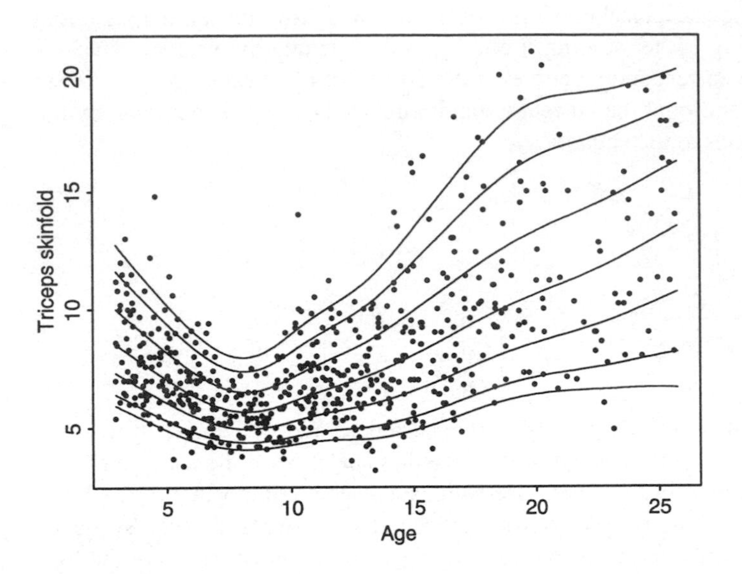

Figure 6.5. *Smoothed reference centile curves for triceps skinfold among Gambian females: 5th, 10th, 25th, 50th, 75th, 90th and 95th percentiles.*

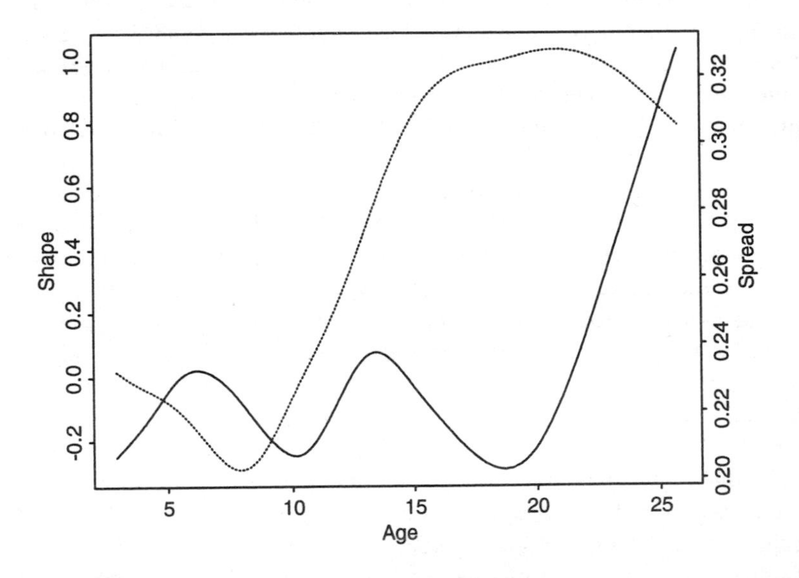

Figure 6.6. *λ (shape, solid line) and σ (spread, broken line) curves for triceps skinfold among Gambian females.*

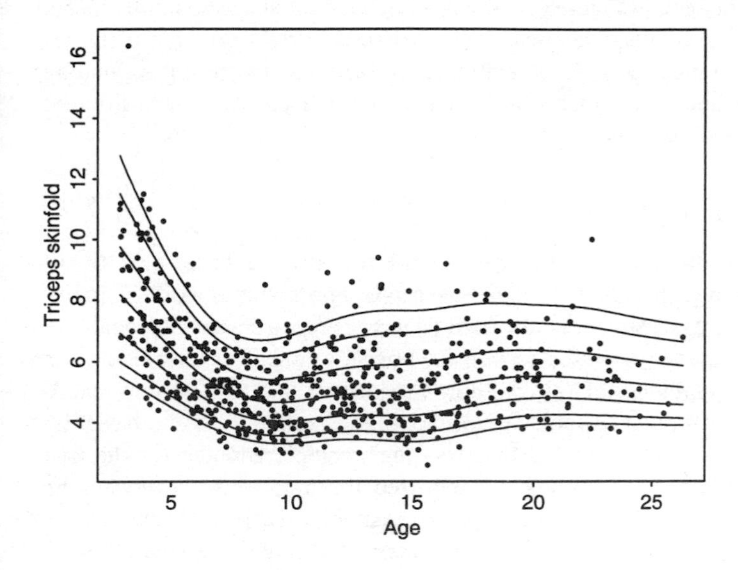

Figure 6.7. *Smoothed reference centile curves for triceps skinfold among Gambian males: 5th, 10th, 25th, 50th, 75th, 90th and 95th percentiles.*

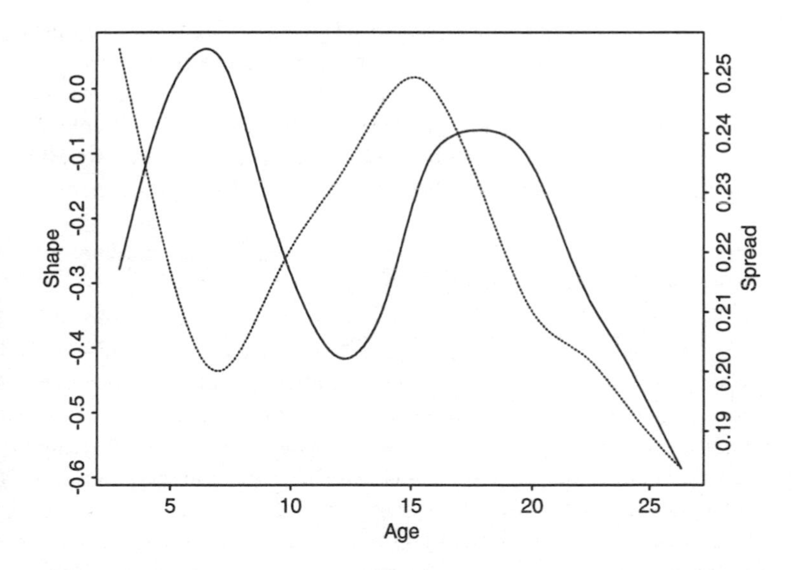

Figure 6.8. λ *(shape, solid line) and* σ *(spread, broken line) curves for triceps skinfold among Gambian males.*

as can be seen in Figure 6.5, the data are somewhat sparse in this range. The relative spread, as measured by $\sigma(t)$, is fairly stable, but increases over the range 10 to 16 years. Figures 6.7 and 6.8 give the corresponding information for the males in the study, and indicate some very different forms of dependence.

6.8 Quasi-likelihood

We close the chapter with an application of the roughness penalty approach to regression that goes beyond likelihood-based models. Sometimes there is insufficient information to construct a complete probability model for the regression responses. For an ordinary linear model, it is of course quite usual to make only second-moment assumptions and to regard least squares rather than maximum likelihood as the basic estimation principle: indeed that was the principle adopted for the first few chapters of this book. Generalized linear models can always be re-parametrized so that the expected responses $\{\mu_i\}$ are treated as the predictors, in the sense of the previous section, and then the different exponential family distributions have characteristic mean/variance relationships mapping $E(Y_i) = \mu_i = b'(\theta_i)$ to $\text{var}(Y_i) = \phi b''(\theta_i)$. For example, the Poisson distribution satisfies $\text{var}(Y_i) = \mu_i$, while the binomial has $\text{var}(Y_i) = \mu_i(1 - \mu_i)/m_i$.

In this spirit, Wedderburn (1974) proposed *quasi-likelihood estimation* in regression models, making use solely of the relationship between the mean and variance of the observations, and the prescribed regression function defining the mean response in terms of explanatory variables. Wedderburn's work was extended by McCullagh (1983) who developed the multivariate case we discuss here, and derived an asymptotic theory for the estimator.

In this section we will describe the corresponding semiparametric version of such models. Suppose that the expected responses are

$$E(Y_i) = \mu_i = \eta_i = \eta_i(\boldsymbol{\beta}, g),$$

that is, we stipulate that the predictors are chosen to coincide with the expected responses. The regression function is subject to the same assumptions as in the previous section: that is, it depends on g only through $g(t_1), \ldots, g(t_n)$. Now suppose that the variance matrix of the responses is

$$\text{var}(\mathbf{Y}) = \phi V(\boldsymbol{\mu})$$

where V is a known matrix function of $\boldsymbol{\mu}$, and ϕ is an unknown scalar. Wedderburn (1974) proposed estimating $\boldsymbol{\beta}$ in parametric models of this form by maximizing the quasi-likelihood Q, defined as any function

$Q\{\mu(\beta)\}$ such that

$$\frac{\partial Q}{\partial \mu} = \phi^{-1} V(\mu)^{-1}(\mathbf{Y} - \mu). \tag{6.13}$$

For motivation, note that this is exactly the form of $\partial \ell / \partial \mu$ in a generalized linear model (see Section 5.2.3): thus the maximum quasi-likelihood estimates in a GLM coincide with the maximum likelihood estimates.

When there is a nonparametric component to the regression function, it is appealing, and directly analogous to our discussion so far, to estimate β and g by maximizing the penalized quasi-likelihood

$$\Pi = Q - \tfrac{1}{2}\lambda \int g''(t)^2 dt.$$

Clearly we have

$$E\left(-\frac{\partial^2 Q}{\partial \mu \partial \mu^T}\right) = \phi^{-1} V^{-1}$$

from (6.13), so that application of Fisher scoring again leads to the estimating equations (5.24) with $\alpha = \lambda\phi$, where now:

$$X = \frac{\partial \mu}{\partial \beta}, \quad N = \frac{\partial \mu}{\partial \mathbf{g}}, \quad W = V^{-1}$$

and

$$\mathbf{z} = (\mathbf{Y} - \mu) + X\beta + N\mathbf{g}.$$

With a linear parametrization of the expectation, $\mu_i = \mathbf{x}_i^T \beta + g(t_i)$, so that $\mathbf{z} = \mathbf{Y}$, and the penalized quasi-likelihood estimates are demonstrably penalized least squares estimates using the weight matrix V^{-1} (and constructed iteratively if this matrix is not a constant).

Within the exponential family, quasi-likelihood estimates coincide with m.l.e.'s, as we stated above: it is of interest to examine the relative efficiency of quasi-likelihood estimates under different sampling distributions (but obtained using the correct mean-variance relationship). This has been done by Firth (1987) who demonstrates that, in parametric regression models, quasi-likelihood estimates retain fairly high efficiency under moderate departures from the appropriate exponential family distribution.

The concept of quasi-likelihood has been extended by Nelder and Pregibon (1987) to allow separate modelling of the dependence of the variance structure of the responses on explanatory variables.

CHAPTER 7

Thin plate splines

7.1 Introduction

In this chapter we shall consider the extension of the penalized least squares idea to regression in more than one dimension. We shall see that there is a natural generalization of smoothing splines to two or more dimensions and that some, but not all, of the attractive features of spline smoothing in one dimension carry over. The method we shall describe is called *thin plate splines*, and for further details the reader is referred to Wahba (1990, Section 2.4) and to the references cited there.

We will mostly be concerned with two-dimensional smoothing, assuming that we have observations at points $\mathbf{t}_1, \mathbf{t}_2, \ldots, \mathbf{t}_n$ of the height of a surface g, where the $\{\mathbf{t}_i\}$ are points in two-dimensional space. The extension to more than two dimensions is discussed in Section 7.9.

We concentrate on interpolation and on penalized least squares regression without covariates, as discussed in Chapter 2. The methods we describe are applicable in the contexts of partial spline fitting and GLM dependence in the obvious way. In the partial spline context, they allow the single splined variable t to be replaced by a vector $\mathbf{t}$ of variables, so that the model fitted is of the form

$$Y_i = \mathbf{x}_i^T \boldsymbol{\beta} + g(\mathbf{t}_i) + \text{ error},$$

where g is now a smooth surface. In the GLM dependence case they again allow the scalar variable t to be replaced by a vector $\mathbf{t}$ throughout.

7.2 Basic definitions and properties

Suppose that we have points $\mathbf{t}_1, \mathbf{t}_2, \ldots, \mathbf{t}_n$ in two-dimensional space $\Re^2$ and that values $g_1, \ldots, g_n$ are given. As in one dimension, the interpolation problem will be that of finding a suitable smooth function g such that $g(\mathbf{t}_i) = g_i$ for $i = 1, 2, \ldots, n$. In this case g will of course be a function of the two-dimensional vector $\mathbf{t}$ for $\mathbf{t}$ in $\Re^2$. The problem of estimating g is thus one of estimating a *surface* rather than a curve as in one dimension.

There is a growing number of fields where estimating surfaces like g is of interest. For example g might be the barometric pressure at a point on the earth's surface, to be reconstructed from observations g_i taken at possibly irregularly spaced weather stations at points $\mathbf{t}_i$. There are, of course, numerous geostatistical problems of other kinds that involve spatial interpolation or smoothing.

In one dimension we posed the problem of finding the smoothest curve g that interpolated a given set of data, defining smoothness in terms of integrated squared second derivative. With this definition of smoothness, the smoothest interpolant was a natural cubic spline with knots at the data points. In order to extend the methodology to the bivariate case, the extension of the definition of smoothness needs to be made, and we discuss this next.

7.2.1 Quantifying the smoothness of a surface

Let us consider a surface g to be *smooth* if it is twice continuously differentiable (though in fact the slightly weaker condition of absolutely continuous first derivatives will suffice). Given any smooth surface g, we wish to define a functional $J(g)$ that measures the overall roughness or 'wiggliness' of g, in an analogous way to the integrated squared second derivative in one dimension. Let us first set out some desiderata for the roughness functional $J(g)$.

1. In some intuitive way, J does indeed measure rapid variation in g and departure from local linearity or flatness.
2. Changing the coordinates by rotation or translation in $\Re^2$ does not affect the value of $J(g)$. This requirement is a very natural one in the spatial context where the coordinate directions and the position of the origin are arbitrary.
3. The functional $J(g)$ is always non-negative, and the class of 'free' functions, those for which $J(g)$ is zero, is a natural one. (Recall that, in the one-dimensional case, functions had zero roughness if and only if they were linear.)
4. The problem of finding the surface g that minimizes $J(g)$ subject to the constraints $g(\mathbf{t}_i) = g_i$ is a tractable one.

In order to define a roughness penalty that satisfies these properties, write (x, y) for the coordinates of a typical point $\mathbf{t}$ in $\Re^2$, so that the function can be written $g(x, y)$. We then set

$$J(g) = \iint_{\Re^2} \left\{ \left(\frac{\partial^2 g}{\partial x^2}\right)^2 + 2\left(\frac{\partial^2 g}{\partial x \partial y}\right)^2 + \left(\frac{\partial^2 g}{\partial y^2}\right)^2 \right\} dx\, dy. \tag{7.1}$$

The penalty function $J(g)$ will be finite provided the second derivatives of g are square-integrable over $\Re^2$. Just as in one dimension this penalty has a mechanical interpretation. Suppose that an infinite elastic flat thin plate is deformed to the shape of the function ϵg for small ϵ. Then the bending energy of the thin plate is to first order proportional to $J(g)$; for this reason the functions that are obtained below, by minimizing $J(g)$ subject to certain constraints, are called *thin plate splines.*

It is clear that $J(g)$ will be large if the function g exhibits high local curvature because this will result in large second derivative; thus J does indeed—in an intuitive sense—quantify the 'wiggliness' of g.

It can be shown, with a little tedious calculus, that if the coordinates in $\Re^2$ are rotated then $J(g)$ is unaffected; the details are left as an exercise for the reader not prepared to take this property on trust.

It is immediate that $J(g)$ is always non-negative, and that $J(g)$ is zero if g is linear. Suppose, conversely, that g is a smooth function with $J(g) = 0$; this implies that the second derivatives of g are all identically equal to zero. From $\frac{\partial^2 g}{\partial x^2} = 0$ we have $g(x, y) = a(y)x + b(y)$ for some functions $a(y)$ and $b(y)$; now use the fact that $\frac{\partial^2 g}{\partial x \partial y} = 0$ to obtain $a'(y) = 0$, so that $a(y) = a$ for some constant a; finally apply $\frac{\partial^2 g}{\partial y^2} = 0$ to conclude that $b''(y) = 0$, so that $b(y) = by + c$ for constants b and c. Putting these calculations together shows that $g(x, y) = ax + by + c$, so that, just as in the one-dimensional case, the roughness penalty $J(g)$ is zero if and only if g is a linear function.

In Sections 7.5 and 7.6 below we shall consider the tractability of the minimization of $J(g)$ subject to interpolation conditions, and the use of $J(g)$ as a roughness penalty in a smoothing procedure. As a preliminary, it is helpful to recast the work on natural cubic splines into a form that is more easily generalized to the multivariate case.

7.3 Natural cubic splines revisited

We saw in Section 2.3 that the function g in $S_2[a, b]$ interpolating $g(t_i) = z_i$ that minimizes $\int g''^2$ is a natural cubic spline. In Section 2.1 two different representations of natural cubic splines were discussed. The most immediate was the piecewise polynomial expression (2.1), where all the coefficients of the various polynomial pieces were given. Of more practical value was the value-second derivative representation set out in Section 2.1.2 in terms of a vector $\mathbf{g}$ of values and γ of second derivatives at the knots. This representation is conceptually simple, and is easily adapted to yield the Reinsch algorithm for the numerical computation of the cubic smoothing spline. However it does not readily generalize to deal with the analogous functions in higher dimensions. As motivation for the more general material later in this chapter, in this section we derive an alternative representation of the natural cubic spline interpolating a given set of data.

Consider the function g defined in terms of constants $a_1, a_2, \delta_1, \delta_2, \ldots, \delta_n$ and $t_1 < t_2 < \ldots < t_n$ by the expression

$$g(t) = a_1 + a_2 t + \frac{1}{12} \sum_{i=1}^{n} \delta_i |t - t_i|^3, \tag{7.2}$$

subject to the constraints

$$\sum_{i=1}^{n} \delta_i = \sum_{i=1}^{n} \delta_i t_i = 0. \tag{7.3}$$

By inspection, this is certainly a cubic spline, with knots at the points $t_1, t_2, \ldots, t_n$; furthermore the constraints (7.3) imply that g'' and g''' are both zero outside $[t_1, t_n]$, so the curve g is a *natural* cubic spline.

Conversely, *any* natural cubic spline g can be written in the form (7.2), subject to (7.3), for it is clear that g can be specified uniquely by its intercept at t_1, $g(t_1) = a_1 + a_2 t_1 + \frac{1}{12} \sum \delta_i (t_i - t_1)^3$, its gradient on $(-\infty, t_1]$, $g'(t) = a_2 - \frac{1}{4} \sum \delta_i t_i^2$, and the increment in its third derivative at each t_i, $g'''(t_i^+) - g'''(t_i^-) = \delta_i$.

Can we find a simple characterization of the a_1, a_2 and $\delta_1, \delta_2, \ldots, \delta_n$ that define the curve g interpolating data (t_i, z_i), and can we evaluate $\int g''^2$ in terms of these coefficients?

Let T be the $2 \times n$ matrix

$$T = \begin{pmatrix} 1 & 1 & \cdots & 1 \\ t_1 & t_2 & \cdots & t_n \end{pmatrix} \tag{7.4}$$

with $T_{1i} = 1$ and $T_{2i} = t_i$ for $i = 1, 2, \ldots, n$, and let E be the $n \times n$ matrix with $E_{ij} = \frac{1}{12}|t_i - t_j|^3$. Then the vector $\mathbf{g}$ of values taken by the curve in (7.2) at the points t_i can be written

$$\mathbf{g} = E\boldsymbol{\delta} + T^T \mathbf{a} \tag{7.5}$$

where $\mathbf{a}$ and $\boldsymbol{\delta}$ are the vectors with components a_i and δ_i. The constraints (7.3) take the form

$$T\boldsymbol{\delta} = 0. \tag{7.6}$$

Thus in our new representation, the natural cubic spline interpolant to the data (t_i, z_i) is found by solving the block matrix equation

$$\begin{bmatrix} E & T^T \\ T & 0 \end{bmatrix} \begin{pmatrix} \boldsymbol{\delta} \\ \mathbf{a} \end{pmatrix} = \begin{pmatrix} \mathbf{z} \\ 0 \end{pmatrix}, \tag{7.7}$$

where $\mathbf{z}$ is the n-vector of values z_i. Just as in the algorithm set out in Section 2.2.1, this requires the solution of a set of linear equations to find the interpolating spline. However it should be noted that the coefficients in (7.7) do not form a band matrix, while those in the previous algorithm have the tridiagonal matrix R of coefficients. Some more remarks about this comparison will be made below.

This deals with the interpolation problem in our new representation, but of course our main interest is in smoothing. To deal with the smoothing problem, we first find an expression for $\int g''^2$. The expression (2.18) shows that, in terms of the value-second derivative representation, the increment in the third derivative of g at t_i

$$\delta_i = \frac{\gamma_{i+1} - \gamma_i}{h_i} - \frac{\gamma_i - \gamma_{i-1}}{h_{i-1}}$$

for $i = 2, \ldots, n-1$. By the definition (2.7) of the matrix Q this yields the matrix equation

$$\boldsymbol{\delta} = Q\boldsymbol{\gamma}. \tag{7.8}$$

But

$$R\boldsymbol{\gamma} = Q^T\mathbf{g} = Q^T(E\boldsymbol{\delta} + T^T\mathbf{a})$$

from (2.4) and (7.5), and since $Q^T T^T$ is easily seen to be 0, we find

$$R\boldsymbol{\gamma} = Q^T E\boldsymbol{\delta}. \tag{7.9}$$

In Section 2.5 we showed that in terms of the value-second derivative representation the integral $\int g''^2$ takes the value $\boldsymbol{\gamma}^T R\boldsymbol{\gamma}$, so using (7.9) and (7.8) we have

$$\int g''(t)^2 dt = \boldsymbol{\gamma}^T Q^T E\boldsymbol{\delta} = \boldsymbol{\delta}^T E\boldsymbol{\delta}. \tag{7.10}$$

We can now express the penalized sum of squares $S(g)$ in terms of our new representation. Given data Y_i as in Section 2.3, we have

$$\begin{aligned} S(g) &= (\mathbf{Y} - \mathbf{g})^T(\mathbf{Y} - \mathbf{g}) + \alpha\boldsymbol{\delta}^T E\boldsymbol{\delta} \\ &= (\mathbf{Y} - E\boldsymbol{\delta} - T^T\mathbf{a})^T(\mathbf{Y} - E\boldsymbol{\delta} - T^T\mathbf{a}) + \alpha\boldsymbol{\delta}^T E\boldsymbol{\delta} \end{aligned} \tag{7.11}$$

where $\boldsymbol{\delta}$ is subject to condition (7.6). The quadratic form (7.11) can be minimized subject to the constraint (7.6) to find $\boldsymbol{\delta}$ and $\mathbf{a}$.

This completes our reworking of the one-dimensional cubic spline smoothing problem. What have we achieved? Just as in the interpolation problem, the numerical linear algebra required in the new representation is less amenable: the constrained minimization of (7.11) does not lead directly to a banded system of equations. The advantages come as we generalize the problem. First note that (7.2) makes no reference to the knots t_i being in increasing numerical order. Secondly, it will be easy to modify (7.2) to represent the solution of penalized least squares problems arising from using different penalty functionals in place of $\int g''^2$, and higher-dimensional t_i.

7.4 Definition of thin plate splines

In this section, we introduce thin plate splines, an important class of functions that will play the rôle that natural cubic splines did in one dimension. Suppose that $\mathbf{t}_1, \mathbf{t}_2, \ldots, \mathbf{t}_n$ are points in $\Re^2$. Before defining a thin plate spline, we need to make some preliminary definitions. Define a function $\eta(r)$ by

$$\begin{aligned} \eta(r) &= \frac{1}{16\pi} r^2 \log r^2 \text{ for } r > 0; \\ \eta(0) &= 0. \end{aligned} \tag{7.12}$$

If a typical point $\mathbf{t}$ has coordinates (x, y), define the three functions ϕ_j on $\Re^2$ by

$$\begin{aligned} \phi_1(x,y) &= 1 \\ \phi_2(x,y) &= x \\ \phi_3(x,y) &= y \end{aligned} \tag{7.13}$$

so that any linear function can be written as a linear combination of the ϕ_j. Define T to be the $3 \times n$ matrix with elements $T_{jk} = \phi_j(\mathbf{t}_k)$, so that

$$T = \begin{bmatrix} 1 & 1 & \cdots & 1 \\ \mathbf{t}_1 & \mathbf{t}_2 & \cdots & \mathbf{t}_n \end{bmatrix}. \tag{7.14}$$

We can now define the two-dimensional analogue of a natural cubic spline. Write $\|\mathbf{t}\|$ for the Euclidean norm of a vector $\mathbf{t}$, $\|\mathbf{t}\|^2 = \mathbf{t}^T\mathbf{t}$.

Definition *A function* $g(\mathbf{t})$ *is a* thin plate spline *on the data set* $\mathbf{t}_1, \ldots, \mathbf{t}_n$ *if and only if g is of the form*

$$g(\mathbf{t}) = \sum_{i=1}^{n} \delta_i \eta(\|\mathbf{t} - \mathbf{t}_i\|) + \sum_{j=1}^{3} a_j \phi_j(\mathbf{t}) \tag{7.15}$$

for suitable constants δ_i *and* a_j. *If the vector* $\boldsymbol{\delta}$ *of coefficients* δ_i *satisfies*

$$T\boldsymbol{\delta} = 0 \tag{7.16}$$

then g is said to be a natural *thin plate spline.*

It is of course immediate that the constraint (7.16) is equivalent to

$$\sum_{i=1}^{n} \delta_i = \sum_{i=1}^{n} \delta_i \mathbf{t}_i = 0$$

corresponding precisely to (7.3). It is convenient in our subsequent discussion to write **a** for the 3-vector of constants a_j. Note that the form of the natural thin plate spline corresponds to the expression (7.2) for a natural cubic spline, with the function $\eta(r)$ replacing $\frac{1}{12}r^3$. Pursuing this correspondence, the $n \times n$ matrix E is defined by

$$\begin{aligned} E_{ij} &= \eta(\|\mathbf{t}_i - \mathbf{t}_j\|) \\ &= \frac{1}{16\pi}\|\mathbf{t}_i - \mathbf{t}_j\|^2 \log \|\mathbf{t}_i - \mathbf{t}_j\|^2 \end{aligned} \tag{7.17}$$

with $E_{ii} = 0$ for each i, making use of the definition of η.

Two important properties of natural thin plate splines are given in the following theorem. The proof of the theorem is left as an exercise for the more mathematically-inclined reader.

Theorem 7.1

1. *If g is a thin plate spline, then $J(g)$ is finite if and only if g is a natural thin plate spline.*
2. *If g is a natural thin plate spline, then*

$$J(g) = \delta^T E \delta. \tag{7.18}$$

The first part of the theorem is an immediate generalization of the fact that, for cubic splines g, $\int_{-\infty}^{\infty} g''^2$ will be infinite unless g is a natural cubic spline. The equation (7.18) corresponds precisely to the expression (7.10) above, and is used in an identical way, as we shall see below.

7.5 Interpolation

7.5.1 Constructing the interpolant

In this section the use of thin plate splines for interpolation will be discussed. Our first result is an exact analogue of Theorem 2.2 in the one-dimensional case.

Theorem 7.2 *Suppose $\mathbf{t}_1, \mathbf{t}_2, \ldots, \mathbf{t}_n$ are distinct non-collinear points in $\Re^2$. Given any values $z_1, z_2, \ldots, z_n$, there is a unique natural thin plate spline g on the set $\mathbf{t}_1, \mathbf{t}_2, \ldots, \mathbf{t}_n$ such that*

$$g(\mathbf{t}_i) = z_i \text{ for } i = 1, \ldots, n. \tag{7.19}$$

Proof. Suppose that g is any thin plate spline satisfying (7.15) above. Then, for each $\mathbf{t}_i$, by the definitions of the matrices E and T,

$$g(\mathbf{t}_i) = \sum_{j=1}^{n} \delta_j \eta(\|\mathbf{t}_i - \mathbf{t}_j\|) + \sum_{k=1}^{3} a_k \phi_k(\mathbf{t}_i) = (E\delta + T^T\mathbf{a})_i, \tag{7.20}$$

so that, exactly as in (7.5), the vector of values taken by g at the points $\mathbf{t}_i$ is $E\delta + T^T\mathbf{a}$. Let $\mathbf{z}$ denote the vector of values z_i. Just as in Section 7.3, it follows from (7.20) and (7.16) that g will be a natural thin plate spline interpolating the values z_i if and only if

$$\begin{bmatrix} E & T^T \\ T & 0 \end{bmatrix} \begin{pmatrix} \delta \\ \mathbf{a} \end{pmatrix} = \begin{pmatrix} \mathbf{z} \\ 0 \end{pmatrix} \tag{7.21}$$

is satisfied.

It will be shown in Lemma 7.1 below that $\begin{bmatrix} E & T^T \\ T & 0 \end{bmatrix}$ is of full rank. It follows at once that the system of equations (7.21) has a unique solution, completing the proof of the theorem. □

The proof of the theorem also yields an algorithm for finding the natural thin plate spline interpolant: set up and solve the equations (7.21).

7.5.2 *An optimality property*

Just as in the one-dimensional case, the natural thin plate spline interpolant has an optimality property, stated in the following theorem.

Theorem 7.3 *The natural thin plate spline interpolant uniquely minimizes $J(g)$ subject to the interpolation conditions $g(\mathbf{t}_i) = z_i$ for all i.*

Proof. We provide a sketch of the proof, which parallels that of Theorem 2.3. Given any smooth functions f and g, let

$$[f, g] = \iint_{\Re^2} \left(\frac{\partial^2 f}{\partial x^2}\frac{\partial^2 g}{\partial x^2} + 2\frac{\partial^2 f}{\partial x \partial y}\frac{\partial^2 g}{\partial x \partial y} + \frac{\partial^2 f}{\partial y^2}\frac{\partial^2 g}{\partial y^2} \right) dx\, dy$$

so that $J(g) = [g, g]$. It can then be shown (Meinguet, 1979) that, if g is a natural thin plate spline on the set $\mathbf{t}_1, \ldots, \mathbf{t}_n$, and if h is any smooth function with $J(h)$ finite, then

$$[g, h] = \sum_{i=1}^{n} \delta_i h(\mathbf{t}_i). \tag{7.22}$$

Now proceed exactly as in the proof of Theorem 2.3; suppose that $\tilde{g}$ is any other interpolant to the values z_i, and set $h = \tilde{g} - g$. It follows at once from (7.22) that $[\tilde{g}, \tilde{g}] = [g, g] + 2[g, h] + [h, h] = J(g) + J(h) \geq J(g)$ with equality only if $J(h) = 0$. If $J(h) = 0$ then h is a linear function, and hence necessarily zero. This completes the proof of the theorem. □

7.5.3 *An example*

In this section we present an example of thin plate spline interpolation, which will also be used in our subsequent development.

Table 7.1. *The positions of 38 data sites and the 'true width' of the ore-bearing layer measured at each site. The coordinates of the data sites are* (t_1, t_2) *and the corresponding measurement is* z

t_1	t_2	z	t_1	t_2	z
–16	–15	17.0	40	4	13.5
–14	–4	18.0	40	–61	18.0
–13	4	17.5	44	–29	19.4
–7	5	19.0	48	–65	13.0
–6	–43	22.0	48	–7	14.0
–6	–36	24.0	49	–32	19.5
1	–50	17.4	55	–71	16.0
2	–39	23.0	56	–14	16.0
2	–8	23.5	59	–38	19.0
2	–51	15.0	62	7	19.0
9	–16	23.5	62	–3	21.5
9	–42	25.0	64	–29	22.0
17	–37	16.5	69	–28	20.5
18	–12	19.5	70	–72	11.0
24	–57	12.0	77	–19	26.0
25	–29	18.5	78	–53	22.0
26	–40	18.0	79	–37	26.0
32	–7	14.0	84	–52	16.0
33	–35	19.0	84	–16	16.0

O'Connor and Leach (1979) present a data set collected from a mine in Cobar, NSW, Australia. At each of 38 sampling points, several measurements were taken. We shall focus attention on one of these, the 'true width' of an ore-bearing rock layer. The coordinates of the data sites, and the value of this observation, are listed in Table 7.1 and the positions of the data points are plotted in Figure 7.1. The convex region Ω shown in the figure has been chosen, for illustrative purposes, to include the data sites reasonably comfortably, without including any regions remote from the convex hull of the data. These data were used by Stone (1988) to illustrate a number of aspects of thin plate spline methods. It can be seen at once that the data points are not regularly spaced in the region. The thin plate spline interpolant to these data is plotted in Figure 7.2. It should be stressed that although the function contours are only shown in the region Ω, the function minimizes the roughness integrated over the whole of $\Re^2$ subject to interpolating the data.

The ability of the method to produce a smooth surface interpolating all the data points is clear. We shall see below that the large-scale structure

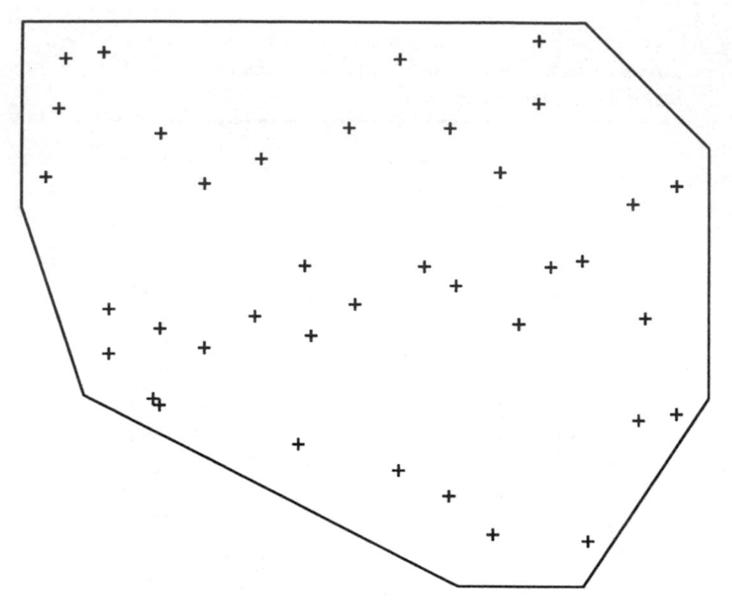

Figure 7.1. *The data sites and a convex window Ω. Reproduced from Stone (1988) with the permission of the author.*

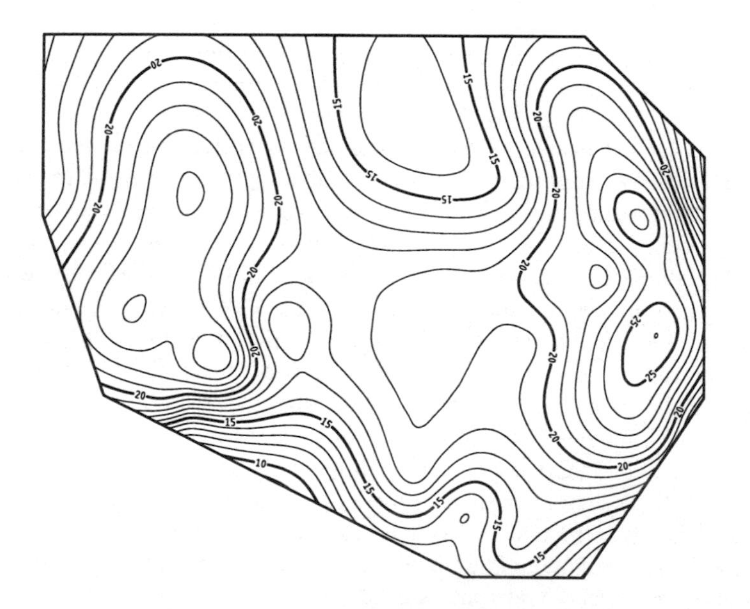

Figure 7.2. *Thin plate spline interpolant. Reproduced from Stone (1988) with the permission of the author.*

of the data becomes clearer if some smoothing is applied, but of course an interpolating surface, however complicated, is important for very many purposes.

It should be noted that the gradient of the thin plate spline interpolant can be worked out explicitly. The gradient of the function g as specified in (7.15) is

$$\nabla g(\mathbf{t}) = \sum_{i=1}^{n} \frac{\delta_i}{8\pi}(\mathbf{t} - \mathbf{t}_i)(1 + \log \|\mathbf{t} - \mathbf{t}_i\|^2) + \begin{pmatrix} a_2 \\ a_3 \end{pmatrix}. \tag{7.23}$$

The contour plot in Figure 7.2 was produced by a contouring routine that makes explicit use of gradient information supplied by the user; because of the explicit formula (7.23) very high quality contour plots can be produced relatively quickly, without the need for an enormous number of evaluations of the interpolant. The contouring routine used for all the figures in this chapter is CONICON3 (Sibson, 1987) which is based on the seamed quadratic element described in Sibson and Thomson (1981).

7.6 Smoothing

7.6.1 Constructing the thin plate spline smoother

Just as in the one-dimensional case, the quantity $J(g)$ can be used as a roughness penalty. Given data values Y_i at the points $\mathbf{t}_i$, we can define the penalized residual sum of squares of a surface g by

$$S(g) = \sum_i \{Y_i - g(\mathbf{t}_i)\}^2 + \alpha J(g). \tag{7.24}$$

Exactly as in one dimension, the parameter $\alpha > 0$ is a smoothing parameter, and the function $S(g)$ combines a term quantifying the lack of fit of g to the data with a roughness penalty term. It follows from (7.24) and Theorem 7.3 that the minimizer of $S(g)$ is necessarily a natural thin plate spline; the argument that demonstrates this is identical with that given in Section 2.3.1 for the one-dimensional case.

Suppose, therefore, that g is a natural thin plate spline. In exactly the same way as (7.11), it follows from (7.20) and from Theorem 7.1 that, letting $\mathbf{Y}$ be the vector with components Y_i,

$$S(g) = (\mathbf{Y} - E\boldsymbol{\delta} - T^T\mathbf{a})^T(\mathbf{Y} - E\boldsymbol{\delta} - T^T\mathbf{a}) + \alpha\boldsymbol{\delta}^T E\boldsymbol{\delta}.$$

We shall now pursue the minimization of this expression for $S(g)$. In matrix form, we have

$$S(g) = \begin{pmatrix} \delta^T & \mathbf{a}^T \end{pmatrix} \begin{bmatrix} E^2 + \alpha E & ET^T \\ TE & TT^T \end{bmatrix} \begin{pmatrix} \delta \\ \mathbf{a} \end{pmatrix} - 2\begin{pmatrix} \delta^T & \mathbf{a}^T \end{pmatrix} \begin{bmatrix} E \\ T \end{bmatrix} \mathbf{Y} + \mathbf{Y}^T\mathbf{Y}. \tag{7.25}$$

Define a natural thin plate spline $\hat{g}$ such that its coefficient vectors satisfy

$$\begin{bmatrix} E + \alpha I & T^T \\ T & 0 \end{bmatrix} \begin{pmatrix} \hat{\delta} \\ \hat{\mathbf{a}} \end{pmatrix} = \begin{pmatrix} \mathbf{Y} \\ 0 \end{pmatrix}. \tag{7.26}$$

It will be shown in Lemma 7.1 below that these equations have a unique solution. Premultiplying the equations (7.26) by the matrix

$$\begin{bmatrix} E & 0 \\ T & -\alpha I \end{bmatrix}$$

yields

$$\begin{bmatrix} E^2 + \alpha E & ET^T \\ TE & TT^T \end{bmatrix} \begin{pmatrix} \hat{\delta} \\ \hat{\mathbf{a}} \end{pmatrix} = \begin{bmatrix} E \\ T \end{bmatrix} \mathbf{Y},$$

which demonstrates that $\hat{\delta}$ and $\hat{\mathbf{a}}$ are the unconstrained minimizers of the expression (7.25), and hence, *a fortiori*, its minimizers subject to the constraint $T\delta$ that ensures that $\hat{g}$ is a natural thin plate spline. Hence it follows that $\hat{g}$ found by solving (7.26) is the unique minimizer of the penalized residual sum of squares $S(g)$.

To sum up, the above discussion demonstrates that the smoothing problem as set out has a unique solution, that can be found by setting up and solving the system of equations (7.26). As in the spatial interpolation case discussed above, this is a full system of linear equations.

7.6.2 An example

To continue the discussion of the example described in Section 7.5.3 above, the thin plate spline smoother $\hat{g}$ applied to the data of Table 7.1 is shown in Figure 7.3. Of course, in principle it is possible to use cross-validation or some other automatic method to choose the smoothing parameter, but for the exploratory purposes of this example the value $\alpha = 10$ was chosen subjectively to display the broad features of the data. It can be seen at once that smoothing makes the overall structure of the data much clearer.

7.6.3 Non-singularity of the defining linear system

In this technical section we demonstrate that the systems of equations (7.21) and (7.26) are non-singular and therefore have unique solutions.

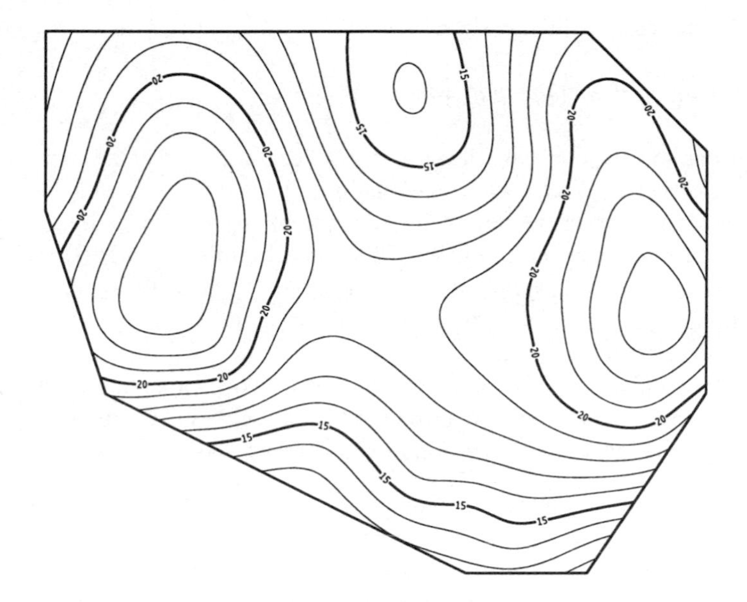

Figure 7.3. *Thin plate spline smoother, with* $\alpha = 10$. *Reproduced from Stone (1988) with the permission of the author.*

The necessary result is the following lemma.

Lemma 7.1 *Suppose that* $t_1, \ldots, t_n$ *are not collinear, and that* $\alpha \geq 0$. *Then the matrix* $\begin{bmatrix} E + \alpha I & T^T \\ T & 0 \end{bmatrix}$ *is of full rank.*

Proof. Suppose that

$$\begin{bmatrix} E + \alpha I & T^T \\ T & 0 \end{bmatrix} \begin{pmatrix} \tilde{\boldsymbol{\delta}} \\ \tilde{\mathbf{a}} \end{pmatrix} = 0,$$

and that $\tilde{g}$ is the thin plate spline defined by the vectors of coefficients $\tilde{\boldsymbol{\delta}}$ and $\tilde{\mathbf{a}}$. Since $T\tilde{\delta} = 0$, $\tilde{g}$ is a natural thin plate spline. Therefore

$$\begin{aligned} J(\tilde{g}) + \alpha \tilde{\delta}^T \tilde{\delta} &= \tilde{\delta}^T (E + \alpha I)\tilde{\delta} = \tilde{\delta}^T (E + \alpha I)\tilde{\delta} + (T\tilde{\delta})^T \tilde{\mathbf{a}} \\ &= \tilde{\delta}^T \{(E + \alpha I)\tilde{\delta} + T^T \tilde{\mathbf{a}}\} = 0. \end{aligned}$$

If $\alpha > 0$ this implies at once that $\tilde{\boldsymbol{\delta}} = 0$ since J only takes non-negative values; if $\alpha = 0$ we can conclude that $J(\tilde{g}) = 0$ so that g is a linear surface, so that $\tilde{\delta}$ again is equal to zero. From $\tilde{\boldsymbol{\delta}} = 0$ it follows that $T^T \tilde{\mathbf{a}} = 0$. Since the $\mathbf{t}_i$ are non-collinear, the rows of T are linearly independent and so this implies that $\tilde{\mathbf{a}} = 0$, completing the proof. □

7.7 Finite window thin plate splines

7.7.1 Formulation of the finite window problem

In the one-dimensional case, it was noted in Section 2.3.4 that the choice of the interval $[a, b]$ over which the roughness penalty is calculated does not essentially affect the interpolation or smoothing splines, as long as all the data points t_i fall into $[a, b]$. The reason for this is that $g''(t) = 0$ if g is a natural cubic spline and t is outside the range of the data, and so extending the interval does not affect the roughness once we restrict attention to natural cubic splines.

In the multivariate case, this is no longer so. Focus attention on the two-dimensional case, and suppose that Ω is a region in $\Re^2$ containing all the data points $\mathbf{t}_i$. Given any smooth functions g and h over Ω, define

$$[g, h]_\Omega = \iint_\Omega (g_{xx}h_{xx} + 2g_{xy}h_{xy} + g_{yy}h_{yy})\, dx\, dy, \qquad (7.27)$$

where the suffices denote partial derivatives with respect to the variables indicated. Define $J_\Omega(g)$ by

$$J_\Omega(g) = [g, g]_\Omega = \iint_\Omega (g_{xx}^2 + 2g_{xy}^2 + g_{yy}^2)\, dx\, dy, \qquad (7.28)$$

so that $J_\Omega(g)$ is the roughness of g restricting attention to the region Ω. If Ω is the whole of $\Re^2$ then J_Ω is precisely the roughness penalty J as defined in (7.1) that forms the basis of the thin plate spline methods. In this section we shall consider the effect of using a finite region, or 'window', Ω instead.

It was pointed out in Section 7.2.1 that $J(g)$ has a physical interpretation, the first order term in the bending energy of an infinite flat thin plate deformed to the shape of the surface g. Using a finite window Ω makes the physical analogy somewhat more realistic, since the energy calculated would be that of a finite thin plate—an object that could actually exist in practice!

The finite window roughness penalty J_Ω can be used in the obvious way, both for interpolation or for smoothing. It is computationally tedious to find the finite window interpolant or smoother precisely, but a good working approximation can be obtained by solving a finite system of linear equations. Some of the mathematical details will be given in Section 7.7.3 below, but first we discuss the effect on the Cobar mine data of taking account of the finite window.

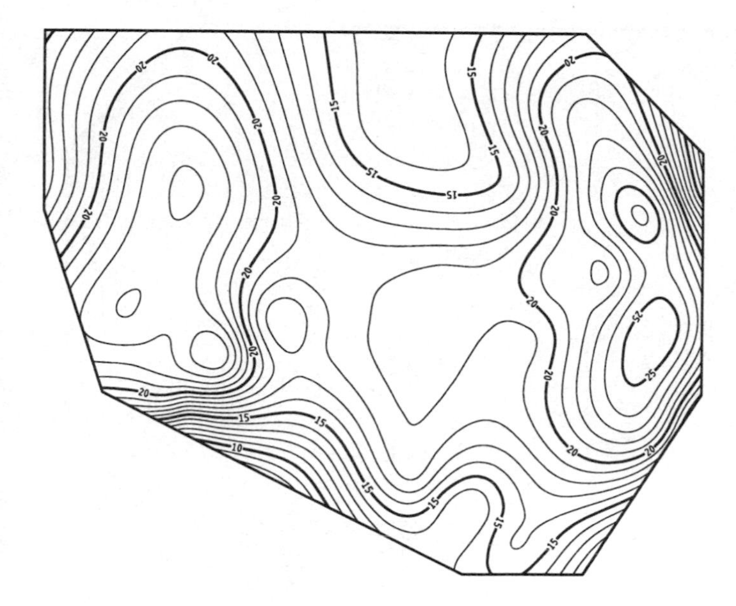

Figure 7.4. *Finite window interpolant. Reproduced from Stone (1988) with the permission of the author.*

7.7.2 An example of finite window interpolation and smoothing

Figure 7.4 shows an approximation to the surface that minimizes $J_{\Omega}(g)$ subject to interpolating the given data. In order to aid understanding of the effect of minimizing $J_{\Omega}(g)$ rather than $J(g)$, we also present in Figure 7.5 the difference between the two interpolants.

It is immediate that, in contrast with the one-dimensional case, restricting to a finite window does affect the interpolant. The main effect is near the edges, as might be expected. In particular, the thin plate spline interpolant tends to prefer closed contours and reduced variability along the edges of the window. The intuitive reason for this is that the thin plate interpolant is forced to tend to a flat function outside the window shown, and that if one drew a very large circle around the data, the values of the interpolant on the circle would have to approximate to those of a linear function in $\mathbf{t}$. Thus long ridges extending outside the range of the data will be discouraged. The finite window interpolant indicates that (for example) the closure of the contour at the middle of the top of the region is not a feature driven by the data, but is a consequence of the tendency of the thin plate spline to abhor variability outside Ω.

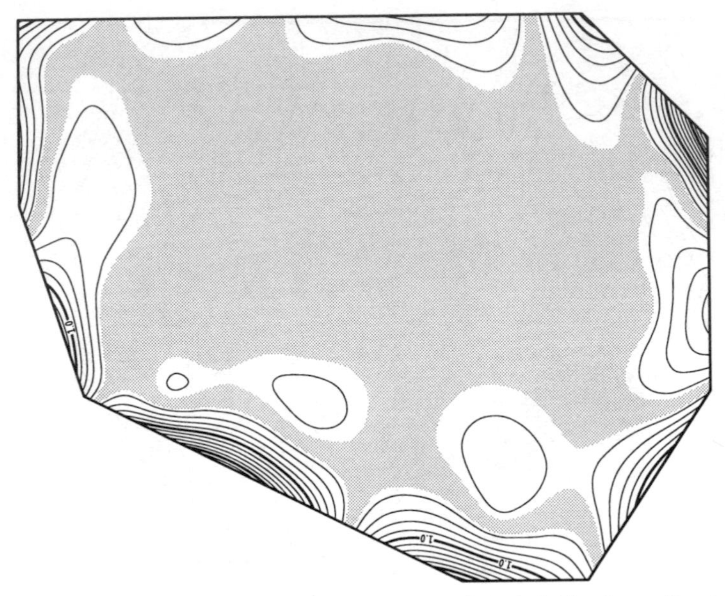

Figure 7.5. *The difference between the finite window and thin plate spline interpolants; the shaded region indicates where the difference is very close to zero. Reproduced from Stone (1988) with the permission of the author.*

We now turn to smoothing. The finite window smoother with smoothing parameter β is, of course, the minimizer $\tilde{g}$ of $\sum\{z_i - g(\mathbf{t}_i)\}^2 + \beta J_\Omega(g)$. It should be noted that it is not immediate how one should compare the two different smoothers. Because a different roughness penalty is used, it is not reasonable merely to set $\beta = \alpha$. A fairer comparison is to choose β so that the residual sum of squares is the same in both cases. The surface corresponding to Figure 7.3 obtained by this approach is shown in Figure 7.6. The differences induced by restricting the roughness penalty to Ω are similar in character (and in fact somewhat larger in magnitude) than those for the interpolation case.

To sum up, this discussion demonstrates that some care is needed in the detailed interpretation of either interpolants or smoothers constructed by the thin plate spline method. In particular one should not place too much credence in the closing of contours in areas of $\Re^2$ not well surrounded by data sites. On the other hand, as long as one does not extrapolate too far, the effect of confining the calculation of the roughness to a finite window is not enormous.

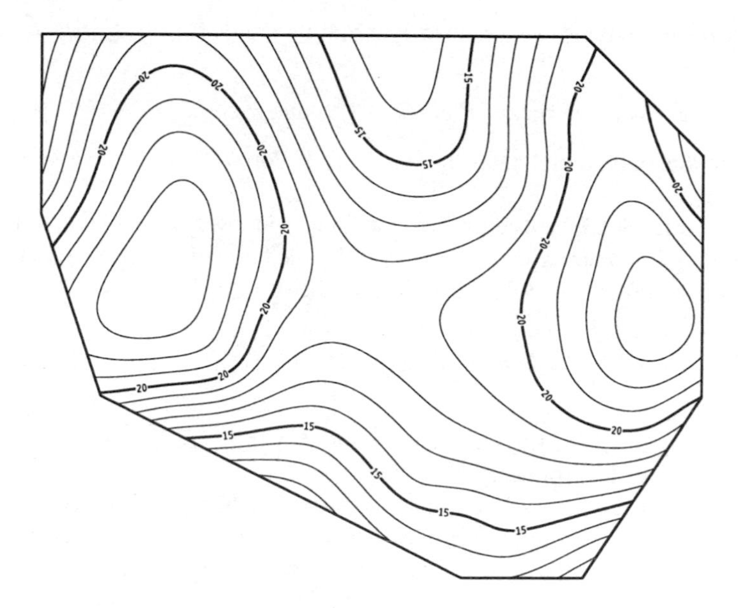

Figure 7.6. *Finite window smoother, chosen to have the same residual sum of squares as Figure 7.3. Reproduced from Stone (1988) with the permission of the author.*

7.7.3 Some mathematical details

In this section, we discuss some of the details of the way in which the finite window interpolant and smoother can be calculated. The detailed description of the algorithm is beyond the scope of this text, and the reader is referred to Dyn and Levin (1982) and Stone (1988) for details.

Define a smooth function ϕ to be *biharmonic* if $\nabla^4\phi(\mathbf{t}) = 0$ for all $\mathbf{t}$. Let $\mathcal{B}$ be the space of biharmonic functions defined on Ω. It can be shown, by a variational argument, that the minimizer of $[g, g]_\Omega$ subject to the interpolation conditions $g(\mathbf{t}_i) = z_i$ for $i = 1, \ldots, n$ can be written as the sum of a thin plate spline and a biharmonic function,

$$g(\mathbf{t}) = \sum_{i=1}^{n} \delta_i \eta(\|\mathbf{t} - \mathbf{t}_i\|) + \phi(\mathbf{t}), \tag{7.29}$$

where the function η is as defined in (7.12) and ϕ is biharmonic. (Note that constant and linear functions are biharmonic, so there is no need to include the second sum in (7.15) explicitly.)

Let $\mathcal{B}_m$ be the space of biharmonic functions that are also polynomials of degree m or less. Let (r, θ) be the polar coordinates of $\mathbf{t}$. It can fairly easily be shown that $\mathcal{B}_m$ is spanned by the polynomials $1, t_1, t_2, t_1^2, t_1t_2, t_2^2$, and, for $k = 3, \ldots, m$,

$$r^k \cos k\theta, \; r^k \sin k\theta, \; r^k \cos(k-2)\theta, \text{ and } r^k \sin(k-2)\theta.$$

Denote this polynomial basis by $\{\phi_1, \ldots, \phi_M\}$. For $m \geq 2$, we have $M = 4m - 2$.

The basic idea of the algorithm for finite window interpolation and smoothing is to approximate the space $\mathcal{B}$ by the finite dimensional space $\mathcal{B}_m$, and therefore to consider functions of the form

$$g(\mathbf{t}) = \sum_{i=1}^{n} \delta_i \eta_i(\mathbf{t}) + \sum_{j=1}^{M} a_j \phi_j(\mathbf{t}). \tag{7.30}$$

where $\eta_i(\mathbf{t}) = \eta(\|\mathbf{t} - \mathbf{t}_i\|)$ for each i. Note that if $m = 1$ then these are precisely thin plate splines.

In principle it is now possible to use the basis functions approach set out in Section 3.6 to find suitable coefficients in (7.30) for interpolation and smoothing. In order to do this, one needs to find $[\eta_i, \eta_j]_\Omega$, $[\eta_i, \phi_j]_\Omega$, and $[\phi_i, \phi_j]_\Omega$, for all relevant i and j. These quantities will yield the components of the $(n+M) \times (n+M)$ matrix K such that the roughness penalty $J_\Omega(g)$ can be written in terms of the coefficient vectors $\boldsymbol{\delta}$ and $\mathbf{a}$ as the quadratic form

$$\begin{pmatrix} \boldsymbol{\delta}^T & \mathbf{a}^T \end{pmatrix} K \begin{pmatrix} \boldsymbol{\delta} \\ \mathbf{a} \end{pmatrix}.$$

The integrals involved in these calculations can be reduced to line integrals by using a standard Green's formula given, for example, by Dyn and Levin (1982). We have

$$[u, v]_\Omega = \int_\Omega v \nabla^4 u + \oint_{\partial\Omega} \{(\nabla v) \cdot \frac{\partial}{\partial n}(\nabla u) - v \frac{\partial}{\partial n}(\nabla^2 u)\}. \tag{7.31}$$

Here $\frac{\partial}{\partial n}$ is the normal derivative at the boundary. If u is a biharmonic function then the integral over Ω of $v\nabla^4 u$ disappears, while if $u = \eta_i$ then $\nabla^4 u$ is a delta function at $\mathbf{t}_i$ so the integral is equal to $v(\mathbf{t}_i)$.

If Ω is a polygonal region then the boundary integral reduces to a sum of integrals along bounded intervals. For this case, Dyn and Levin (1982) and Stone (1988) describe ingenious computational approaches to the calculation of the boundary integral in the case where u is one of the basis biharmonic polynomials ϕ_j. They also set out a further approximation in the basis function approach that can be used to avoid calculating the quantities $[\eta_i, \eta_j]_\Omega$.

Stone (1988) investigated the degree m of the biharmonic polynomials space $\mathcal{B}_m$ that should be used in practice. He found that setting $m = 7$ (so that there are 26 basis functions ϕ_i) gave very satisfactory results. In the examples he considered, including the one presented in Section 7.7.2, it was clear that most of the finite window effect was accounted for using this value of m. The effect of increasing m still further (to 9 or 10) was perceptible but unimportant.

7.8 Tensor product splines

Another possible approach to the problem of smoothing over a finite window is via the use of *tensor product splines*, which are a systematic method of using families of smooth functions in one dimension to generate smooth surfaces in higher dimensional spaces. They provide an approach rather different from thin plate splines, but we shall see that there are circumstances in which tensor product splines can be used to approximate thin plate spline smoothers. Tensor product splines are widely used in many areas of approximation theory and numerical analysis, and we only give a brief treatment here. As in our discussion of thin plate splines, we shall assume that we are fitting data by a model

$$Y = g(\mathbf{t}) + \text{error}$$

where the vector $\mathbf{t}$ is higher-dimensional. One of our motivations for the discussion of tensor product splines is that they can be used as a convenient basis for the approximation of finite window roughness penalty smoothers using the basis functions approach of Section 3.6.

7.8.1 *Constructing tensor products of one-dimensional families*

For ease of exposition, we will just consider the *two*-dimensional case, and model smooth functions defined on a rectangle $T \times U$ in $\Re^2$, where T and U are intervals in $\Re$. Given one-dimensional functions $\delta : T \to \Re$ and $\epsilon : U \to \Re$, the tensor product of δ and ϵ is the function $\delta \otimes \epsilon : T \times U \to \Re$ defined by $(\delta \otimes \epsilon)(t, u) = \delta(t)\epsilon(u)$. Suppose we have a set of linearly independent functions $\{\delta_{j_1} : j_1 = 1, 2, \ldots, q_1\}$ defined on T that we wish to regard as smooth, and similarly a basis $\{\epsilon_{j_2} : j_2 = 1, 2, \ldots, q_2\}$ of smooth functions on U. The tensor product of these two spaces of functions is the set of all linear combinations of tensor products of linear combinations of these basis functions, that is

$$\mathcal{G} = \left\{ \sum_r c_r \left(\sum_{j_1=1}^{q_1} a_{rj_1} \delta_{j_1} \right) \otimes \left(\sum_{j_2=1}^{q_2} b_{rj_2} \epsilon_{j_2} \right) \right\}.$$

This is the same as the set of all linear combinations of tensor products of basis functions

$$\mathcal{G} = \left\{ \sum_{j_1=1}^{q_1} \sum_{j_2=1}^{q_2} g_j \delta_{j_1} \otimes \epsilon_{j_2} \right\}. \tag{7.32}$$

In this section, we use a subscript j to denote the pair (j_1, j_2), and later use k similarly. The $q_1 q_2$ functions $\delta_{j_1} \otimes \epsilon_{j_2}$ are easily shown to be linearly independent and to form a basis for $\mathcal{G}$.

Tensor product cubic splines are an obvious example of this construction. Suppose $\{\tau_1, \tau_2, \ldots, \tau_{m_1}\}$ and $\{\upsilon_1, \upsilon_2, \ldots, \upsilon_{m_2}\}$ are regularly spaced sequences such that $T = [\tau_1, \tau_{m_1}]$ and $U = [\upsilon_1, \upsilon_{m_2}]$. Then the cubic splines on each of these knot sequences, restricted to the intervals T and U respectively, are finite-dimensional spaces of dimension $q_1 = m_1 + 2$ and $q_2 = m_2 + 2$. A function in $\mathcal{G}$ can be defined by splitting $T \times U$ into rectangular *panels*, small rectangles of the form $[\tau_r, \tau_{r+1}] \times [\upsilon_s, \upsilon_{s+1}]$. Over each panel the function is the product of a cubic in t and a cubic in u, and the functions fit together smoothly (continuous first and second derivatives) at the joins between the panels. The characterization (7.32) allows us to express a generic tensor product spline in terms of any convenient bases for the one-dimensional cubic splines, and we shall give a specific example in the next subsection.

7.8.2 *A basis function approach to finite window roughness penalties*

Given a particular tensor product space $\mathcal{G}$, it is within the space $\mathcal{G}$ that we shall seek a smooth function to fit data supposed to follow the model

$$Y_i = g(t_i, u_i) + \text{error} \quad \text{for } i = 1, 2, \ldots, n.$$

Note that in contrast to the thin plate splines approach as described above, this is not an isotropic treatment of the regression problem, as the space $\mathcal{G}$ is not invariant to rotations of the (t, u) space. Nevertheless, we can still use isotropic roughness penalties within the space $\mathcal{G}$.

A natural example is the thin plate penalty $J_{T \times U}(g)$ defined as in equation (7.28), the finite window roughness penalty with the range of integration restricted to the rectangle $T \times U$. Suppose that $g = \sum_j g_j \delta_{j_1} \otimes \epsilon_{j_2}$, and define the inner product $[\cdot, \cdot]_{T \times U}$ as in (7.27). Let K be the $q_1 q_2 \times q_1 q_2$ matrix defined by

$$K_{jk} = [\delta_{j_1} \otimes \epsilon_{j_2}, \delta_{k_1} \otimes \epsilon_{k_2}]_{T \times U} = \sum_{s=0}^{2} \binom{2}{s} D_{j_1 k_1}^{(s)} E_{j_2 k_2}^{(2-s)} \tag{7.33}$$

where

$$D^{(s)}_{j_1k_1} = \int_T \delta^{(s)}_{j_1}(t)\delta^{(s)}_{k_1}(t)dt \text{ for } j_1, k_1 = 1, 2, \ldots, q_1,$$

$$E^{(s)}_{j_2k_2} = \int_U \epsilon^{(s)}_{j_2}(u)\epsilon^{(s)}_{k_2}(u)du \text{ for } j_2, k_2 = 1, 2, \ldots, q_2.$$

Then

$$\begin{aligned} J_{T\times U}(g) &= [\sum_j g_j\delta_{j_1} \otimes \epsilon_{j_2}, \sum_k g_k\delta_{k_1} \otimes \epsilon_{k_2}]_{T\times U} \\ &= \sum_j\sum_k g_jg_k[\delta_{j_1} \otimes \epsilon_{j_2}, \delta_{k_1} \otimes \epsilon_{k_2}]_{T\times U} = \mathbf{g}^T K\mathbf{g}, \end{aligned}$$

where $\mathbf{g}$ is the q_1q_2-vector with entries g_j.

In the present context, the penalized sum of squares $S_W(g)$ of (3.21) takes the form

$$\begin{aligned} S_W(g) &= \sum_{i=1}^n w_i\{Y_i - g(t_i, u_i)\}^2 + \alpha J(g) && (7.34) \\ &= (\mathbf{Y} - N\mathbf{g})^T W(\mathbf{Y} - N\mathbf{g}) + \alpha\mathbf{g}^T K\mathbf{g}, && (7.35) \end{aligned}$$

where N is the $n \times q_1q_2$ matrix with $N_{ij} = \delta_{j_1}(t_i)\epsilon_{j_2}(u_i)$. The value of $\mathbf{g}$ minimizing $S_W(g)$ is therefore, once again,

$$(N^T WN + \alpha K)^{-1}N^T W\mathbf{Y}.$$

The key simplifying feature of these calculations is that, because of the expression (7.33) and the definition of N, the matrices K and N can be built up from properties of the one-dimensional bases considered separately.

Let us now be more specific, and, following work by Inoue (1986), work out some details of this procedure for particular bases $\{\delta_{j_1}\}$ and $\{\epsilon_{j_2}\}$. Let $\{\tau_r\}$ and $\{\upsilon_s\}$ be regular sequences as above, such that if $T = [\tau_1, \tau_{m_1}]$ and $U = [\upsilon_1, \upsilon_{m_2}]$ then the rectangle $T \times U$ includes all observed (t_i, u_i). We will use cubic splines, with knots at the $\{\tau_r\}$ and $\{\upsilon_s\}$. Let β denote the cubic B-spline given by

$$\beta(x) = \begin{cases} B_1(x+3) & \text{for} \quad -3 \le x \le -2 \\ B_2(x+2) & \text{for} \quad -2 \le x \le -1 \\ B_3(x+1) & \text{for} \quad -1 \le x \le 0 \\ B_4(x) & \text{for} \quad 0 \le x \le 1 \\ 0 & \text{otherwise} \end{cases}$$

where

$$\begin{aligned}
B_1(x) &= x^3/6 \\
B_2(x) &= (-3x^3+3x^2+3x+1)/6 \\
B_3(x) &= (3x^3-6x^2+4)/6 \\
B_4(x) &= (1-x)^3/6.
\end{aligned}$$

Let $\delta = \tau_2 - \tau_1$ and $\epsilon = \upsilon_2 - \upsilon_1$ be the knot spacings. We now define our basis functions by

$$\begin{aligned}
\delta_{j_1}(t) &= \beta\left(\frac{t-\tau_{j_1}}{\delta}\right) \\
\epsilon_{j_2}(u) &= \beta\left(\frac{u-\upsilon_{j_2}}{\epsilon}\right)
\end{aligned}$$

for $j_1 = 1, 2, \ldots, m_1 + 2$ and $j_2 = 1, 2, \ldots, m_2 + 2$.

In each dimension, the basis functions are equally spaced translates of each other. This leads to some particularly simple expressions: for example any function $g \in \mathcal{G}$ can be written

$$\begin{aligned}
g(t,u) &= \sum_j g_j \delta_{j_1}(t)\epsilon_{j_2}(u) \\
&= \sum_{j_1=r}^{r+3}\sum_{j_2=s}^{s+3} g_j B_{4+r-j_1}\left(\frac{t-\tau_r}{\delta}\right) B_{4+s-j_2}\left(\frac{u-\upsilon_s}{\epsilon}\right)
\end{aligned}$$

if $(t,u) \in [\tau_r, \tau_{r+1}] \times [\upsilon_s, \upsilon_{s+1}]$. The entries in the matrix N are readily shown to be

$$\begin{aligned}
N_{ij} &= \delta_{j_1}(t_i)\epsilon_{j_2}(u_i) \\
&= B_{4+r-j_1}\left(\frac{t_i-\tau_r}{\delta}\right) B_{4+s-j_2}\left(\frac{u_i-\upsilon_s}{\epsilon}\right)
\end{aligned}$$

if $(t_i, u_i) \in [\tau_r, \tau_{r+1}] \times [\upsilon_s, \upsilon_{s+1}]$, provided $r \le j_1 \le r+3$ and $s \le j_2 \le s+3$, and 0 otherwise.

Finally, the integrated derivative matrices are

$$\begin{aligned}
D^{(s)}_{j_1k_1} &= \int_T \delta^{(s)}_{j_1}(t)\delta^{(s)}_{k_1}(t)dt \\
&= \delta^{1-2s}\sum_r \int_0^1 B^{(s)}_{r-j_1}(x)B^{(s)}_{r-k_1}(x)dx
\end{aligned}$$

where the sum runs from $r = \max\{5, j_1+1, k_1+1\}$ to $\min\{m_1+3, j_1+4, k_1+4\}$, with a similar expression for $E^{(s)}_{j_2k_2}$. Thus, because of the finite support $[-3, 1]$ for β, all these matrices have some banded structure. Specifically, K is a sum of Kronecker (or outer) products of matrices with bandwidth 7, and (N^TWN) likewise has the property that its (j,k)

entry is 0 unless both $|j_1 - k_1| \leq 3$ and $|j_2 - k_2| \leq 3$. Also, except at the boundary, $D^{(s)}_{j_1 k_1}$ depends only on $k_1 - j_1$, and similarly for $E^{(s)}$.

7.9 Higher order roughness functionals

Our development so far has been based on use of $\int g''^2$ to measure the roughness of curves g in one dimension, and the functional $J(g)$ as defined in (7.1) to quantify the roughness of a two-dimensional surface. These simple devices have taken us a long way. The idea of smoothness implicit in the use of these penalties, based on second derivatives, corresponds to that appreciated visually, and the penalties lead to interpolants and smoothers with good properties. However, for some purposes it is useful to make use of higher derivatives than the second in measuring roughness. Further, it is certainly of interest to extend the applicability of the roughness functional idea to functions g of more than two variables.

Consider a function g of a variable $\mathbf{t}$ in $\Re^d$. Among the important properties of the penalties $\int g''^2$ and $J(g)$ were their invariance under translations and rotations of the coordinate system, and their quadratic dependence on g, which led to fitted curves and surfaces linear in the data values z_i or Y_i. A penalty in d dimensions based on mth derivatives that retains these properties is

$$J_m(g) = \int \cdots \int_{\Re^d} \sum \frac{m!}{\nu_1! \ldots \nu_d!} \left(\frac{\partial^m g}{\partial t_1^{\nu_1} \ldots \partial t_d^{\nu_d}} \right)^2 dt_1 \ldots dt_d, \tag{7.36}$$

where the sum within the integral is now over all non-negative integers $\nu_1, \nu_2, \ldots, \nu_d$ such that $\nu_1 + \ldots + \nu_d = m$. This most general form that we will consider remains non-negative and isotropic, and has the property that the only surfaces for which $J(g)$ is zero are the polynomials of total degree less than m.

In order to make progress, it is necessary to impose the condition $2m > d$, so that roughness functionals J_m based on integrated first derivatives can be used only in one dimension, those based on integrated second derivatives only in three or fewer dimensions, and so on. The reason for this restriction can be expressed technically in terms of Beppo Levi and Sobolev spaces; see, for example, Meinguet (1979). A relatively non-technical explanation can be given by considering the following example. Suppose that g is a smooth unimodal radially-symmetric 'bump' function, with $J_m(g)$ finite, such that $g(\mathbf{0}) = 1$ and $g(\mathbf{t})$ is zero outside the unit disc. Now define $g_n(\mathbf{t})$ to be the function $g(n\mathbf{t})$, so that as n increases g_n approaches a 'spike' of height 1 at the origin. Under any reasonable definition of roughness, we would require that the roughness of g_n increases as n increases. A simple argument involving the change of variables $\mathbf{s} = n\mathbf{t}$ in the definition of J_m shows that

$$J_m(g_n) = n^{2m-d} J_m(g),$$

so that if $2m \leq d$ the limiting spike will have finite roughness, and indeed if $2m < d$ the roughness of the sequence g_n will actually tend to zero as n increases!

7.9.1 Higher order thin plate splines

We can now address the problem of fitting a smooth surface g to data from the model

$$Y_i = g(\mathbf{t}_i) + \text{ error },$$

for $i = 1, 2, \ldots, n$, where $\mathbf{t}_i$ is a d-vector. The estimator of g that we will consider is, as usual, the minimizer of a penalized sum of squares

$$S_{md}(g) = \sum_{i=1}^{n} \{Y_i - g(\mathbf{t}_i)\}^2 + \alpha J_m(g),$$

where $J_m(g)$ is as defined in (7.36).

Yet again, we can treat this problem by dealing first with interpolation and by restricting attention to a finite-dimensional class of functions g.

Define a function η_{md} by

$$\eta_{md}(r) = \begin{cases} \theta\, r^{2m-d} \log r & \text{if } d \text{ is even} \\ \theta\, r^{2m-d} & \text{if } d \text{ is odd} \end{cases} \tag{7.37}$$

where the constant of proportionality θ is given by

$$\theta = \begin{cases} (-1)^{m+1+\frac{d}{2}} 2^{1-2m} \pi^{-\frac{d}{2}} (m-1)!^{-1} (m - \frac{d}{2})!^{-1} & \text{for } d \text{ even} \\ \Gamma(\frac{d}{2} - m) 2^{-2m} \pi^{-\frac{d}{2}} (m-1)!^{-1} & \text{for } d \text{ odd.} \end{cases}$$

Let $M = \binom{m+d-1}{d}$, and let $\{\phi_j, j = 1, 2, \ldots, M\}$ be linearly independent polynomials spanning the M-dimensional space of polynomials in $\Re^d$ of total degree less than m.

Define a function g on $\Re^d$ to be a *natural thin plate spline of order m* if g is of the form

$$g(\mathbf{t}) = \sum_{i=1}^{n} \delta_i \eta_{md}(\|\mathbf{t} - \mathbf{t}_i\|) + \sum_{j=1}^{M} a_j \phi_j(\mathbf{t}), \tag{7.38}$$

and the vector of coefficients δ satisfies the condition

$$T\delta = 0, \tag{7.39}$$

where T is the $t \times n$ matrix with

$$T_{ij} = \phi_i(\mathbf{t}_j). \tag{7.40}$$

Duchon (1976) and Meinguet (1979) showed that, provided the points $\mathbf{t}_i$ are distinct and sufficiently dispersed to determine a unique least squares polynomial surface of total degree $m - 1$, and that $2m > d$, the function g minimizing $J_m(g)$ subject to $g(\mathbf{t}_i) = z_i$ is a natural thin plate spline of order m. By exactly the same arguments as in Section 7.5, the vectors $\boldsymbol{\delta}$ and $\mathbf{a}$ of coefficients in (7.38) satisfy the matrix equation (7.21), defining the matrix E by

$$E_{ij} = \eta_{md}(\|\mathbf{t}_i - \mathbf{t}_j\|). \tag{7.41}$$

and the matrix T as in (7.40).

The discussion of Section 7.6 carries over immediately to the minimization of S_{md}. The minimizing function is a natural thin plate spline of order m, with coefficient vectors uniquely specified by

$$\begin{bmatrix} E + \alpha I & T^T \\ T & 0 \end{bmatrix} \begin{pmatrix} \boldsymbol{\delta} \\ \mathbf{a} \end{pmatrix} = \begin{pmatrix} \mathbf{Y} \\ 0 \end{pmatrix},$$

where T and E are as defined in (7.40) and (7.41) above.

It is easily checked from the definitions that if g is a natural thin plate spline of order m in $\Re^d$ then g will have $2m - d - 1$ continuous derivatives everywhere, but its $(2m - d)$th derivative has a singularity at each data point $\mathbf{t}_i$. Thus, in one dimension for example, the use of the penalty $\int g'^2$ leads to piecewise linear interpolants and 'smoothers', whose derivatives jump at each data point; more generally, in odd dimensions the use of a penalty with $2m = d+1$ will yield a function with a singularity resembling the point of a cone at each data point. If a visibly smooth function is required it is therefore advisable to use a penalty with $2m > d + 1$.

In the case $d = 1$ the discussion of this section shows that the penalty $\int g^{(m)}(t)^2 dt$ will lead to splines that are piecewise $(2m - 1)$th degree polynomials. There are algorithms corresponding to the Reinsch algorithm for finding these higher order smoothing splines in $O(n)$ operations. For details, see, for example, De Boor (1978).

CHAPTER 8

Available software

In this chapter we give some details of software available at the time of writing for the techniques described in this book. There are many implementations other than those discussed in detail here. For example, a FORTRAN program for the Reinsch algorithm described in Section 2.3.3 is given by DeBoor (1978, Chapter 14); see also the IMSL routine ICSSCU.

8.1 Routines within the S language

8.1.1 The S routine `smooth.spline`

The statistical language S* and its extended version S-PLUS include a routine called `smooth.spline` which can be used for one-dimensional cubic spline smoothing as discussed in Chapters 1, 2 and 3. There is no requirement that the design points $\{t_i\}$ be distinct or ordered.

The routine allows the use of weights, and thus implements, in addition, the general smoothing operation that we have denoted by

$$S = N(N^T WN + \alpha K)^{-1} N^T W$$

where N is an incidence matrix, W a diagonal weight matrix, and K the kernel of the roughness penalty for cubic spline smoothing. This smoother was encountered first in Section 3.5, was an important building block in implementing the fitting of partial spline models in 4.3, and appeared in the algorithms involving Fisher scoring in 5.3 and 5.4. The routine uses the basis function algorithm (Section 3.6) with a basis of B-splines.

The value of the smoothing parameter in the routine is specified with respect to units in which the $\{t_i\}$ are rescaled to have range 1 and the weights (if any) rescaled to sum to n, the number of data points. To translate to 'raw' values, the value α_S of the smoothing parameter within

* Commercially available from StatSci, 1700 Westlake Ave. N., Suite 500, Seattle, WA 98109, USA.

the routine corresponds to the minimization of

$$\sum_{i=1}^{n} w_i\{Y_i - g(t_i)\}^2 + \alpha \int g''^2 \tag{8.1}$$

with

$$\alpha = \{(\max t_j - \min t_j)^3 n^{-1} \sum_{i=1}^{n} w_i\}\, \alpha_S.$$

The amount of smoothing can be specified either in terms of the smoothing parameter α_S or in terms of the equivalent degrees of freedom, as defined in Section 3.3.4. This latter feature may well provide a more intuitive way of specifying the 'complexity' of the fitted curve. If desired, an automatic choice of smoothing can be made either by cross-validation or by generalized cross-validation (Section 3.3).

If there are fewer than 50 distinct points $\{t_i\}$, the routine minimizes (8.1) over cubic splines with knots allowed at all the data points, so the curve found is precisely the cubic spline smoother. For larger data sets, the default—which can be overridden—is only to allow knots at a particular subset of the $\{t_i\}$. The size of this subset increases slowly as n increases. This feature makes for economy in time and storage while only having a perceptible effect on the result for very small values of the smoothing parameter.

For further details of the routine, see Chambers and Hastie (1992), consult the S documentation, or—for online help within an S session—type `?smooth.spline` or `help(smooth.spline)`. The package also provides an auxiliary routine `predict.smooth.spline`, which allows the calculation of the estimated curve or its derivatives at any specified points.

8.1.2 The new S modelling functions

As well as the usual updates and enhancements, the 1991 release of S incorporated a new system for regression model fitting, including facilities for linear regression, analysis of variance, generalized linear models, tree-based models, and nonlinear regression. The functions that fit additive and generalized additive models are of particular interest in the context of this book. All the modelling routines adopt common conventions about data handling, and specification of models, and thus S provides a system of considerable power for those who wish to fit regression models flexibly, without the labour of programming.

Full details of the philosophy and use of these new facilities is given by Chambers and Hastie (1992). Here we will just give a brief taste of the possibilities by noting that the model we proposed for the marketing example in Section 4.5 can be specified by calling the function

```
gam(log(volume) ~ price+diffprice+day+s(date))
```

where `volume`, `price`, `diffprice`, `day` and `date` are variables and factors with the obvious interpretation. By changing the specification of the nonparametric part of the model to `s(date,df=18.7)`, the degree of smoothing applied would be the same as used in Figure 4.1. However, `gam()` uses backfitting, and as we observed in Section 4.5, this algorithm converges unacceptably slowly for these data.

Similarly, the model used for the tumour data in Section 5.5 can be specified and fitted by

```
gam(lesion ~ log(dose)+weight+position+s(age),
                family=binomial).
```

8.2 The GCVPACK routines

The package GCVPACK (Bates *et al.*, 1987) is a useful set of subroutines written in FORTRAN 77, designed around the themes of thin plate spline smoothing and generalized cross-validation. They are available free of charge from Netlib.* The routines can be used for smoothing in any number of dimensions, including the one-dimensional case, and allow penalty functions involving any number of derivatives. In the one-dimensional case, however, they do not take advantage of the special structure that allows $O(n)$ calculations, and so they run more slowly than a good implementation of the Reinsch and Hutchison–de Hoog algorithms, or of a suitable basis function algorithm.

Facilities are also provided for partial splines with one or more splined variables, and for more general linear smoothing problems. In each case the package allows the choice of smoothing parameter by generalized cross-validation or the use of a smoothing parameter value specified by the user. The package makes extensive use of the LINPACK linear algebra library (Dongarra *et al.*, 1979).There are three main driver routines, each of which is a FORTRAN 77 program with a large number of arguments, allowing considerable flexibility.

The GCVPACK routines can also be used as building blocks for an implementation of the approaches to generalized linear models set out in Chapter 5. For example, the package PGLMPACK (Yandell, 1988) carries out generalized cross-validation calculations in a range of semiparametric

* Information about Netlib can be obtained by sending an email message to `netlib@research.att.com` with the single line `send index`. The routines are in the directory `gcv`.

generalized linear models. Models with binomial, Poisson and normal dependence are catered for within the package. Any number of splined variables can be used, with an appropriate thin plate spline roughness penalty.

8.2.1 Details of the algorithm

The key technique used by GCVPACK is the singular value decomposition, together with other, related, matrix decompositions. In this subsection, we set out the algorithm used for thin plate spline smoothing, and refer the reader to Bates *et al.* (1987) for details of the extensions to partial splines and other models. Except where otherwise stated, our notation will be the same as that used in Chapter 7. It will be shown that the GCV choice of smoothing parameter, and the calculation of the estimate itself, becomes very economical once certain matrix decompositions have been carried out. For details of the various matrix decompositions, see, for example the LINPACK manual (Dongarra *et al.*, 1979.)

As explained in Sections 7.3, 7.6 and 7.9, the thin plate spline smoother can be found by solving a system of equations

$$\begin{bmatrix} E+\alpha I & T^T \\ T & 0 \end{bmatrix} \begin{pmatrix} \boldsymbol{\delta} \\ \mathbf{a} \end{pmatrix} = \begin{pmatrix} \mathbf{Y} \\ 0 \end{pmatrix}, \tag{8.2}$$

where $\boldsymbol{\delta}$ and $\mathbf{a}$ are the coefficients of the minimizing thin plate spline and E and T are matrices that depend on the design points $\mathbf{t}_i$ and on the roughness penalty being used.

The first step is to perform a QR decomposition of T^T, to obtain

$$T^T = FG$$

where F is an $n \times n$ orthogonal matrix and G is $n \times t$ with zeroes below the main diagonal. Let F_1 be the first t columns of F and F_2 the last $n-t$ columns, and let G_1 be the first t rows of G; since G is upper triangular, its remaining rows are zero and hence $T^T = F_1 G_1$.

Assuming that the rows of T are linearly independent, G_1 will be non-singular and hence $T\boldsymbol{\delta} = 0$ if and only if $F_1^T \boldsymbol{\delta} = 0$. By the orthogonality of F, this implies that $\boldsymbol{\delta} = F_2 \boldsymbol{\zeta}$, where $\boldsymbol{\zeta}$ is the $(n-t)$-vector $F_2^T \boldsymbol{\delta}$. Set $\mathbf{w}_1 = F_1^T \mathbf{Y}$ and $\mathbf{w}_2 = F_2^T \mathbf{Y}$. Then multiplying the first block of (8.2) by F_1^T and F_2^T respectively leads, after some manipulation, to

$$G_1 \mathbf{a} = \mathbf{w}_1 - F_1^T E \boldsymbol{\delta} \tag{8.3}$$

and

$$(F_2^T E F_2 + \alpha I)\boldsymbol{\zeta} = \mathbf{w}_2. \tag{8.4}$$

Now decompose $F_2^T E F_2$ in the form UD^2U^T, where U is an orthogonal $(n-t)\times(n-t)$ matrix, and D is a diagonal matrix with diagonal entries $d_1 \geq d_2 \geq \ldots \geq d_{n-t} \geq 0$. In the GCVPACK package, this is done by Cholesky decomposition of $F_2^T E F_2$ as LL^T, with L lower triangular, followed by singular value decomposition of L as $L = UDV^T$. Bates and Wahba (1982) point out that the computational burden here can be alleviated, at the expense of approximation, by truncating the singular value decomposition.

No further decompositions are needed: the entire computation of the smoother can now be expressed via transformation of the problem using F and U, and scalar operations using the eigenvalues d_j^2. First, equation (8.4) is rearranged to yield

$$\zeta = U(D^2+\alpha I)^{-1}U^T\mathbf{w}_2. \tag{8.5}$$

A convenient expression for the generalized cross-validation score can now be derived. The vector of predicted values is equal to

$$\begin{aligned}
E\delta + T^T\mathbf{a} &= F_2F_2^TEF_2\zeta + F_1F_1^TE\delta + F_1G_1\mathbf{a}\\
&= F_2(\mathbf{w}_2-\alpha\zeta)+F_1\mathbf{w}_1\\
&= F_2\{I-\alpha U(D^2+\alpha I)^{-1}U^T\}\mathbf{w}_2 + F_1\mathbf{w}_1\\
&= \{F_2UD^2(D^2+\alpha I)^{-1}U^TF_2^T + F_1F_1^T\}\mathbf{Y}.
\end{aligned}$$

Thus the hat matrix $A(\alpha)$ can be written

$$F\begin{bmatrix} I & 0\\ 0 & U\end{bmatrix}\begin{bmatrix} I & 0\\ 0 & D^2(D^2+\alpha I)^{-1}\end{bmatrix}\begin{bmatrix} I & 0\\ 0 & U^T\end{bmatrix}F^T, \tag{8.6}$$

where I is the identity $t\times t$ matrix in each case. Since F and U are orthogonal, and D is diagonal, the trace of the hat matrix is $t+\sum_j\{d_j^2/(d_j^2+\alpha)\} = n-\alpha\sum_j(d_j^2+\alpha)^{-1}$.

To write the residual sum of squares in a convenient form, set $\mathbf{z} = U^T\mathbf{w}_2$. From (8.6) and the fact that F and U are orthogonal matrices, it follows that $I-A(\alpha)$ is equal to

$$F\begin{bmatrix} I & 0\\ 0 & U\end{bmatrix}\begin{bmatrix} 0 & 0\\ 0 & \alpha(D^2+\alpha I)^{-1}\end{bmatrix}\begin{bmatrix} I & 0\\ 0 & U^T\end{bmatrix}F^T,$$

and hence the residual sum of squares $\|\{I-A(\alpha)\}\mathbf{Y}\|^2$ is equal to $\|\alpha(D^2+\alpha I)^{-1}\mathbf{z}\|^2$. Substituting the expressions for the residual sum of squares and the trace of the hat matrix into the definition (3.13) of GCV, we obtain the generalized cross-validation score

$$GCV(\alpha) = n\frac{\sum_{j=1}^{n-t}(d_j^2+\alpha)^{-2}z_j^2}{\{\sum_{j=1}^{n-t}(d_j^2+\alpha)^{-1}\}^2}.$$

Once the necessary decompositions have been carried out to find the d_j and z_j, it is a very economical matter to minimize this expression to find the automatic choice of α. It then follows from (8.5) that $\boldsymbol{\delta} = F_2U(D^2 + \alpha I)^{-1}\mathbf{z}$, again an easy calculation. Finally, (8.3) can be used to find $\mathbf{a}$, again a straightforward calculation since G_1 is a triangular matrix.

References

Ahlberg, J. H., Nilson, E. N. and Walsh, J. (1967) *The Theory of Splines and their Applications*. New York: Academic Press.

Barndorff-Nielsen, O. E. and Blæsild, P. (1986) A note on the calculation of Bartlett adjustments. *J. Roy. Statist. Soc.* B, **48**, 353–358.

Bates, D. M., Lindstrom, M. J., Wahba, G. and Yandell, B. S. (1987) GCVPACK — Routines for generalized cross validation. *Commun. Statist. Simul. Comput.*, **16**, 263–297 (Algorithms Section).

Bates, D. M. and Wahba, G. (1982) Computational methods for generalized cross validation with large data sets. In *Treatment of Integral Equations by Numerical Methods* (C. T. H. Baker and G. F. Miller, eds). New York: Academic Press.

Becker, R. A., Chambers, J. M. and Wilks, A. R. (1988) *The New S Language: A Programming Environment for Data Analysis and Graphics*. Pacific Grove, California: Wadsworth & Brooks.

Besag, J. E. and Kempton, R. A. (1986) Statistical analysis of field experiments using neighboring plots. *Biometrics*, **42**, 231–251.

Breiman, L. and Friedman, J. H. (1985) Estimating optimal transformations for multiple regression and correlation (with discussion). *J. Amer. Statist. Assoc.*, **80**, 580–619.

Buckley, M. J. and Eagleson, G. K. (1988) An approximation to the distribution of quadratic forms in normal random variables. *Austral. J. Statist.*, **30A**, 150–159.

Buckley, M. J., Eagleson, G. K. and Silverman, B. W. (1988) The estimation of residual variance in nonparametric regression. *Biometrika*, **75**, 183–199.

Buja, A., Hastie, T. J. and Tibshirani, R. J. (1989) Linear smoothers and additive models (with discussion). *Ann. Statist.*, **17**, 453–555.

Carter, C. K., Eagleson, G. K. and Silverman, B. W. (1992) A comparison of the Reinsch and Speckman splines. *Biometrika*, **79**, 81–91.

Chambers, J. M. and Hastie, T. J. (1992) *Statistical Models in S.* Pacific Grove, California: Wadsworth & Brooks.

Chu, C.-K. and Marron, J. S. (1992) Choosing a kernel regression estimator (with discussion and rejoinder). *Statistical Science*, **6**, 404–436.

Cole, T. J. (1988) Fitting smoothed centile curves to reference data (with discussion). *J. Roy. Statist. Soc.* A, **151**, 385–418.

Cole, T. J. and Green, P. J. (1992) Smoothing reference centile curves: the LMS method and penalized likelihood. *Statistics in Medicine*, **11**, 1305–1319.

Cook, R. D. and Weisberg, S. (1982) *Residuals and Influence in Regression*. London: Chapman and Hall.

Cordeiro, G. M. (1985) Improved likelihood ratio statistics for generalized linear models. In *Generalized Linear Models* (R. Gilchrist, B. J. Francis and J. Whittaker, eds), Lecture Notes in Statistics, **32**. Berlin: Springer.

Cox, D. D. and Koh, E. (1986) A smoothing spline based test of model adequacy in polynomial regression. Technical Report 787, Department of Statistics, University of Wisconsin–Madison.

Cox, D. D., Koh, E., Wahba, G. and Yandell, B. S. (1988) Testing the (parametric) null model hypothesis in (semiparametric) partial and generalized spline models. *Ann. Statist.*, **16**, 113–119.

Craven, P. and Wahba, G. (1979) Smoothing noisy data with spline functions. *Numer. Math.*, **31**, 377–390.

Cuzick, J. (1992) Semiparametric additive regression. *J. Roy. Statist. Soc.* B, **54**, 831–843.

Daniel, C. and Wood, F. S. (1980) *Fitting Equations to Data: Computer Analysis of Multifactor Data for Scientists and Engineers*. 2nd Edition. New York: Wiley.

De Boor, C. (1978) *A Practical Guide to Splines*. New York: Springer-Verlag.

Dinse, G. E. and Lagakos, S. W. (1983) Regression analysis of tumour prevalence data. *Applied Statistics*, **32**, 236–248.

Dinse, G. E. and Lagakos, S. W. (1984) Correction to Dinse and Lagakos (1983). *Applied Statistics*, **33**, 79–80.

Dongarra, J. J., Bunch, J. R., Moler, C. B. and Stewart, G. W. (1979) *Linpack Users' Guide*. Philadelphia: SIAM.

Duchon, J. (1976) Interpolation des fonctions de deux variables suivant le principe de la flexion des plaques minces. *R.A.I.R.O. Analyse numérique*, **10**, 12, 5–12.

Dyn, N. and Levin, D. (1982) Construction of surface spline interpolants of scattered data over finite domains. *R.A.I.R.O. Analyse numérique*, **16**, 3, 201–209.

Engle, R. F., Granger, C. W. J., Rice, J. A. and Weiss, A. (1986) Semiparametric estimates of the relation between weather and electricity sales. *J. Amer. Statist. Assoc.*, **81**, 310–320.

Eubank, R. L. (1984) The hat matrix for smoothing splines. *Statist. Prob. Letters*, **2**, 9–14.

Eubank, R. L. (1985) Diagnostics for smoothing splines. *J. Roy. Statist. Soc.* B, **47**, 332–341.

Eubank, R. L. (1988) *Spline Smoothing and Nonparametric Regression*. New York: Marcel Dekker.

Finney, D. J. (1947) *Probit Analysis*. Cambridge: Cambridge University Press.

Firth, D. (1987) On the efficiency of quasi-likelihood estimation. *Biometrika*, **74**, 233-245.

Gasser, T., Sroka, L. and Jenner, C. (1986) Residual variance and residual pattern in nonlinear regression. *Biometrika*, **73**, 625–633.

Golub, G. H. and Van Loan, C. F. (1983) *Matrix Computations*. Baltimore: Johns Hopkins University Press.

Good, I. J. and Gaskins, R. A. (1971) Nonparametric roughness penalties for probability densities. *Biometrika*, **58**, 255–277.

Good, I. J. and Gaskins, R. A. (1980) Density estimation and bump-hunting by the penalized maximum likelihood method exemplified by scattering and meteorite data. *J. Amer. Statist. Assoc.*, **75**, 42–73.

Green, P. J. (1984) Iterated reweighted least squares for maximum likelihood estimation, and some robust and resistant alternatives (with discussion). *J. Roy. Statist. Soc.* B, **46**, 149–192.

Green, P. J. (1985) Linear models for field trials, smoothing and cross-validation. *Biometrika*, **72**, 523–537.

Green, P. J. (1987) Penalized likelihood for generalized semi-parametric regression models. *Int. Statist. Rev.*, **55**, 245–259.

Green, P. J. (1989) Generalised linear models and some extensions: geometry and algorithms. *Statistical Modelling* (A. Decarli, B. J. Francis, R. Gilchrist and G. U. H. Seeber, eds), Lecture Notes in Statistics, **57**, pp. 26–36. Berlin: Springer.

Green, P. J., Jennison, C., and Seheult, A. (1983) Contribution to the discussion of Wilkinson *et al.* (1983), *J. Roy. Statist. Soc.* B, **45**, 193–195.

Green, P. J., Jennison, C., and Seheult, A. (1985) Analysis of field experiments by least squares smoothing. *J. Roy. Statist. Soc.* B, **47**, 299–315.

Green, P. J. and Yandell, B. (1985) Semi-parametric generalized linear models. *Generalized Linear Models* (R. Gilchrist, B. J. Francis and J. Whittaker, eds), Lecture Notes in Statistics, **32**, pp. 44–55. Berlin: Springer.

Gu, C. (1992) Cross-validating non-Gaussian data. *Journal of Computational and Graphical Statistics*, **1**, 169–179.

Härdle, W. (1990) *Applied Nonparametric Regression*. Cambridge: Cambridge University Press.

Hastie, T. J. and Tibshirani, R. J. (1990) *Generalized Additive Models*. London: Chapman and Hall.

Hastie, T. J. and Tibshirani, R. J. (1993) Varying-coefficient models (with discussion). *J. Roy. Statist. Soc.* B, **55**, 757–796.

Heckman, N. E. (1986) Spline smoothing in a partly linear model. *J. Roy. Statist. Soc.* B, **48**, 244–248.

Hutchison, M. F. and de Hoog, F. R. (1985) Smoothing noisy data with spline functions. *Numer. Math.*, **47**, 99–106.

Inoue, H. (1986) A least-squares smooth fitting for irregularly spaced data: finite-element approach using the cubic B-spline basis. *Geophysics*, **51**, 2051–2066.

Jørgensen, B. (1983) Maximum likelihood estimation and large-sample inference for generalized linear and nonlinear regression models. *Biometrika*, **70**, 19–28.

Kimeldorf, G. and Wahba, G. (1970) A correspondence between Bayesian estimation on stochastic processes and smoothing by splines. *Ann. Math. Statist.*, **41**, 495–502.

London, D. (1985) *Graduation: the Revision of Estimates*. Winsted, Connecticut: ACTEX Publications.

McCullagh, P. (1983) Quasi-likelihood functions. *Ann. Statist.*, **11**, 59–67.

McCullagh, P. and Nelder, J. A. (1989) *Generalized Linear Models*. 2nd edition. London: Chapman and Hall.

Meinguet, J. (1979) Multivariate interpolation at arbitrary points made simple. *ZAMP*, **30**, 292–304.

Nelder, J. A. and Pregibon, D. (1987) An extended quasi-likelihood function. *Biometrika*, **74**, 221–232.

Nelder, J. A. and Wedderburn, R. W. M. (1972) Generalized linear models. *J. Roy. Statist. Soc.* A, **135**, 370–384.

O'Connor, D. P. H. and Leach, B. G. (1979) Geostatistical analysis of 18CC Stope block, CSA mine, Cobar, NSW. *Estimation and Statement of Mineral Reserves*, pp. 145–153. Melbourne: Australian IMM.

Osborne, C. (1990) *Non-parametric Calibration*. PhD Thesis, University of Bath.

Osborne, C. (1991) Statistical calibration: A review. *Int. Statist. Rev.*, **59**, 309–336.

O'Sullivan, F., Yandell, B. S. and Raynor, W. J. (1986) Automatic smoothing of regression functions in generalized linear models. *J. Amer. Statist. Assoc.*, **81**, 96–103.

Prenter, P. (1975) *Splines and Variational Methods*. New York: Wiley.

Ramsay, C. M. (1993) Minimum variance moving-weighted-average graduation. *Trans. Soc. Actuaries*, **43**, 21–41.

Reinsch, C. (1967) Smoothing by spline functions. *Numer. Math.*, **10**, 177–183.

Rice, J. (1984) Bandwidth choice for nonparametric regression. *Ann. Statist.*, **12**, 1215–1230.

Rice, J. (1986) Convergence rates for partially splined models. *Statist. Prob. Letters*, **4**, 203–208.

Rosenblatt, M. (1991) *Stochastic Curve Estimation*. NSF-CBMS Regional Conference Series in Probability and Statistics, Volume 3. Hayward, California: Institute of Mathematical Statistics.

Scallan, A., Gilchrist, R. and Green, M. (1984) Fitting parametric link functions in generalised linear models. *Comp. Statist. and Data Anal.*, **2**, 37–49.

Schumaker, L. L. (1993) *Spline Functions: Basic Theory*. Melbourne, Florida: Krieger.

Sibson, R. (1987) CONICON3 contouring package. University of Bath.

Sibson, R. and Thomson, G. D. (1981) A seamed quadratic element for contouring. *Computer Journal*, **24**, 378–382.

Silverman, B. W. (1978) Density ratios, empirical likelihood and cot death. *Appl. Statist.*, **27**, 26–33.

Silverman, B. W. (1984a) Spline smoothing: the equivalent variable kernel method. *Ann. Statist.*, **12**, 898–916.

Silverman, B. W. (1984b) A fast and efficient cross-validation method for smoothing parameter choice in spline regression. *J. Amer. Statist. Assoc.*, **79**, 584–589.

Silverman, B. W. (1985) Some aspects of the spline smoothing approach to nonparametric regression curve fitting (with discussion). *J. Roy. Statist. Soc.* B, **47**, 1–52.

Silverman, B. W. (1992) Should we use kernel methods at all? *Statistical Science*, **6**, 430–433.

Silverman, B. W. and Wood, J. T. (1987) The nonparametric estimation of branching curves. *J. Amer. Statist. Assoc.*, **82**, 551–558.

Speckman, P. (1988) Kernel smoothing in partial linear models. *J. Roy. Statist. Soc.* B, **50**, 413–436.

Steer, B. T. and Hocking, R. A. (1985) The optimum timing of nitrogen application to irrigated sunflowers. *Proceedings of the Eleventh International Sunflower Conference, Mar del Plata, Argentina*, pp. 221–226. Buenos Aires: Asociacion Argetina de Girasol.

Stone, G. (1988) *Bivariate Splines*. PhD thesis, University of Bath.

Todd, J. (1962) *Survey of Numerical Analysis*. New York: McGraw-Hill.

Tukey, J. W. (1977) *Exploratory Data Analysis*. Reading, MA: Addison-Wesley.

Utreras, F. (1980) Sur le choix du parametre d'ajustement dans le lissage par fonctions spline. *Numer. Math.*, **34**, 15–28.

Wahba, G. (1978) Improper priors, spline smoothing, and the problem of guarding against model errors in regression. *J. Roy. Statist. Soc.* B, **40**, 364–372.

Wahba, G. (1983) Bayesian confidence intervals for the cross-validated smoothing spline. *J. Roy. Statist. Soc.* B, **45**, 133–150.

Wahba, G. (1990) *Spline Models for Observational Data*. Philadelphia: SIAM.

Wedderburn, R. W. M. (1974) Quasi-likelihood functions, generalised linear models, and the Gauss-Newton method. *Biometrika*, **61**, 439–447.

Wedderburn, R. W. M. (1976) On the existence and uniqueness of maximum likelihood estimates for certain generalized linear models. *Biometrika*, **63**, 27–32.

Whittaker, E. (1923) On a new method of graduation. *Proc. Edinburgh Math. Soc.*, **41**, 63–75.

Whittle, P. (1985) Proposal of the vote of thanks to Silverman (1985). *J. Roy. Statist. Soc.* B, **47**, 21–22.

Wilkinson, G. N., Eckert, S. R., Hancock, T. W. and Mayo, O. (1983) Nearest neighbour (NN) analysis of field experiments (with discussion). *J. Roy. Statist. Soc.* B, **45**, 151–211.

Williams, E. R. (1985) A neighbour model for field experiments. *Biometrika*, **73**, 279–287.

Yandell, B. S. (1988) Algorithms for multidimensional semiparametric GLM's. *Commun. Statist.* B, **17**, 295–312.

Author index

Subject index